ARE WE WORTH OUR SALT?

"Great Spirits have always encountered violent opposition from mediocre minds."

Albert Einstein

ARE WE WORTH OUR SALT?

A Simple Look at Earth Changes

Past Results and Future Implications

Arthur Ryan

ISBN 978-0-6152-1270-8

Schiller-David Publishing

450 Main St.

Circle Pines, MN 55014

This Book is dedicated to my wife, Wendy

for her patience.

I hope it pays off!

I wish to acknowledge Steve Battis, Dwayne "Smokey" Iverson, and Walt Marz for their hours of insightful conversation that helped formulate the ideas discussed in this book. Without their input these words would be locked away in my head forever.

Excerpt from Foreword to "The Earth's Shifting Crust" by Charles H. Hapgood in cooperation with James H. Campbell

"In a polar region there is a continual deposition of ice, which is not symmetrically distributed about the pole. The Earth's rotation acts on these unsymmetrically deposited masses {of ice}, and produces centrifugal momentum that is transmitted to the rigid crust of the Earth. The constantly increasing centrifugal momentum produced in this way will, when it has reached a certain point, produce a movement of the Earth's crust over the rest of the Earth's body, and this will displace the polar regions towards the equator."

Albert Einstein

Contents

FOREWORD

Like many of you, I spent the majority of my early life in classrooms listening to teachers and reading books as they attempted to explain the answers to our World's mysteries. I was too young and ignorant to challenge what I was hearing. I simply believed it all. After all, I had little reason to doubt their ideas since I had no knowledge of any other possibilities. To make matters worse, our education system was, and continues to be based upon right or wrong answers, in the eyes of the instructor. There is no platform for debate in our classrooms. Sadly, challenge to what is being taught is frowned upon as causing disruption.

This process began in elementary school, and soon followed me into middle school, later into high school and finally several years of college where I studied mathematics, biology, chemistry, quantum physics and thermodynamics while earning an Engineering degree. The whole time I continued focusing more on my grades than on the truth.

As my life progressed, I got busier and busier. Even now, as I pursue the completion of my Masters degree, I find that I am too busy to stop and reconsider other possibilities to the answers I received for the riddles I never solved. Was I falling victim to the same "dumbing down" process that has plagued Mankind for so many centuries? I simply stopped asking "Why?" Yet, I continue to find discomfort in what I learned to be "the truth" as if my subconscious recognizes that the answers I've been given, just do not add up.

Even with all my education in the sciences, I do not claim to be an expert in any specific subject such as Archeology, Meteorology, Physics, Geology or History. But I do share something common in all of us: *That burning desire to understand where we came from, and where we are going.*

This book served to awaken my yearning to understand the puzzles that we find throughout our fascinating World; a yearning that has been buried by my ignorance for many years. The most important lesson that our author, Arthur Ryan has taught me; something I failed to do in my youth as a student: was to question everything, to not believe all that you hear, while at the same time, to not be too quick to dismiss new ideas.

In my search for the truth, I have discovered that I need to be able to hold multiple ideas in my head at the same time. The goal should never be to quickly reach a solution, as we've always been taught, but rather to continually and relentlessly challenge <u>and</u> consider all possibilities.

Whether or not the demise of our current epoch will be the result of causes exposed within this book, I believe the underlining message that Art is trying to communicate is that we at least begin taking a more serious approach toward understanding the World around us. The unique thing about Art Ryan is that he is not interested in justifying grant money spent on a focused line of study, such as Geology or History, nor is he interested in being a hero that uncovers the Holy

Grail of one of Earth's head-scratching mysteries.

Through our many discussions, I have come to know Art as more like a detective who studies bits and pieces from observations made throughout the World, and throughout time. These bits and pieces are forged together into a giant puzzle using all the sciences as resources in his quest for the truth. A process which lead him to believe that Mankind must stop working in a vacuum, and begin to work collectively, without the bias toward their own field of study, to debate and consider the real truth about our World.

I believe we all share in the responsibility to ensure the survival of our species. We owe it to those who gave their lives in the past, and to those who depend upon us for their mere existence in the future. We need to try to understand the risks facing our future existence and work toward minimizing those risks with a sense of urgency.

Many religions believe that the end of civilization is near. This may cause multitudes to become complacent in their quest for survival. But no one really knows exactly how or when our World will change. Should we not do all we can to protect ourselves against natural disasters as we grow to understand each of them? Can we afford to ignore our basic instinct for survival? And if we cannot prevent certain inevitable catastrophes, should we not at least consider a system of documentation and communication that would survive the elements of earth and time, so that future civilizations can learn of our discoveries and be better equipped to pick up where we left off?

I hope as each of you discovers the multitude of unanswered mysteries exposed in this book and explore the possible explanations as consideration for what might have happened as well as what may occur. I believe you will, as I have, come away with a new perspective. You'll find yourself asking "What if?" or "Why not?" And during that process of transformation, I hope you will come away as leaders willing to awaken the rest of our World to new ideas. And more importantly, to begin the dialog that needs to take place if we are to ever seriously consider surviving our future.

If we do not do something, and do something now – who will?!

Steve Battis

Preface

The hardest thing about writing this book was finding a place to start. Topics tended to expand and even wander away from the main idea. Yet, if you spend years exploring the implications of each piece of the puzzle, the importance of their inclusion seems critical. This is one such book.

Some many years back, my father, Ben, sent me a book to read while I was deployed in the military. The note attached simply said "I think you will find this interesting." In my mind, no more inadequate statement was ever made.

The book - "The HAB Theory" by Allan W. Eckert told a story of the days prior to the cataclysmic event of a pole-shift. The story was compelling and seemed to address many anomalies that the sciences seem to ignore. It was a fiction book. I later learned that it was written to draw attention to non-fiction book on the same subject. It was called "Cataclysms of the Earth" by Hugh Auchinloss Brown. These books used the premise that the entire earth fell over on its side as a result of a build-up of ice on the polar caps.

Now you must know that I am just an average, blue collar, American - who attended high school, joined the military, took some college courses, and later settled into a job that was able to keep my interest for awhile. I am not an expert on any of the subjects that will be addressed, even though my writing style may give the impression that I think I am. I am not a want-a-be scientist or anything like it.

I am just a person who had doubts about things that were pressed into our minds as fact. I looked at things with a critical eye because I just could not understand how things could be the way we were taught.

The spark caused by my exposure to "The HAB Theory", sent me on a quest for knowledge that spans over two decades. This mental journey began as a search for proof that the theory was wrong. As I traveled the various paths that I came upon, I found only additional evidence that it is true. Much of the items can be found in the aforementioned works with additional twists added in "Path of the Poles" by Charles H. Hapgood, who narrowed the scope of the flip to just the outer crust shifting.

The journey also ventured into areas that, at first look, appear unrelated. This, however, was not to be true. Whether you look at the topics as evidence of this theory or as hints to larger theories, in my mind they all apply.

Please remember that I am but a simple man trying to express an idea. I offer these writings to open some minds to the possibilities. I understand I do not hold the credentials that will command automatic respect for my hypothesis. Don't let that stop you from reading this book. The original theory received positive words of encouragement from Albert Einstein. The ideas expounded upon within this and the other cited books are simple and don't require a person to believe everything. Just look at them and question the current dogma.

Questioning whether what is taught to us is the only answer, thinking outside of the box, and reviewing those things that I present is exactly what the purpose of this book is. Doubt is the fuel that drives new ideas and findings. So, don't take my word for it, open your mind, read on, and most of all – I ask that you look at everything thing you think of as fact with an eye of doubt.

Finding the correct answers is a must!

"I frequently receive communications from people who wish to consult me concerning their unpublished ideas. It goes without saying that these ideas are very seldom possessed of scientific validity. The very first communication, however, that I received from Mr. Hapgood electrified me."

Albert Einstein

Chapter One

How Come?

Unanswered Anomalies

In my years of researching an apparent crustal shift, I came upon a number of anomalies. These weird items were out there; but, had not been given an explanation by the sciences. They seem to be ignored. As if people were afraid to come up with an answer that either defeated their weakly supported theory or more importantly it, the answer, would point to something that they did not want to be true.

Protecting their theories is understandable when you consider that there is only so much research money available. To go out on a limb with a new idea could cause grant money to dry up. Very few scientists wish to risk their livelihood to express an alternate theory that would not find support among their peers. Hungry scientists are not heard from long. They become ostracized and drummed out or they capitulate and return to the accepted dogma.

Others may follow the evidence and find that the path moves in a direction that may be currently associated with the lunatic fringe. No self-respecting scientist wants his/her colleagues to think of them as crazy. After all, apocalyptic events are only addressed in myth or religion. Science has better things to do. The end of civilization in the "here-and-now" is not one of them. Future events far from our present are OK, but not tomorrow or the next day.

It has also occurred to me that our educational process has created a large number of experts in a specific discipline that fail to look at items that traditionally belong to another science. Geologists are geologists and archeologists are archeologists. Thus, no one is crossing boundaries to look for evidence that confirms they are right or wrong in their own theories or they drone along by following the accepted dogma.

As a result, I became aware that in order to look at abnormalities, you need not be an extremely educated person and you surely cannot be relying on grant money or prestige within a discipline. Only free thinkers, who have nothing to lose can question the status quo. An Average Joe. So, here I go!

Why are there ancient savannas under the North African deserts?

How come the largest herbivore dinosaurs found to date were discovered in the desert?

How did aquatic animals end up fossilized in the mountains?

What would cause wooly mammoths to become frozen in permafrost and how did they remain that way protecting them from decay and scavengers?

Why is there so much evidence around the planet that supports a flood theory?

I think this is enough questions for now. More will arise and will be addressed. Even more interestingly; a great number will not be covered in this book, yet they exist. If I miss some, try applying this theoretical idea to them and see how they pan out. Will this book give the answers or will the additional questions provide an opportunity to prove this incorrect. The truth is within our grasp.

Back to the questions that have been asked herein. Let's give them some time.

Savannas under deserts. Evidently a major climate change occurred. A drastic change, that if it took place over the span of many years, would have caused a blending of strata layers as remaining evidence. There would not be a definite line of change between two strata. The savanna would have slowly burned away under the extreme weather. This burning over time should have actually left nearly no evidence of its existence. A form of decomposing would have taken place and transformed the flora into a dirt-like substance. In other words, the change would have to have taken place relatively quickly to leave evidence of the savanna's flora. Otherwise, the intense light (heat) would have burnt away the remnants of the plant life that had been there. Remember the lessons taught us in school about the Dust Bowl Period in American History. While not a direct example, it can be used to visualize the process of change. So, in order for the savannas to have left evidence of their existence in the North Africa regions prior to the desert, then a very sudden change in climate is required.

<u>Dinosaurs in the desert</u>. Without question dinosaurs

did not and could not live in the desert. They called the savannas or other previous environments their home. They probably thrived in lush greenery found there. Yet, somehow they stayed even though the climate was changing for the worst, that's if you agree with the current dogma. I doubt it! Next, the belief would be that they died and were somehow preserved before the change in climate. Again, I don't think the natural instinct of scavengers and/or oxidation stopped; so, I doubt it! The survival instinct of these animals would have caused them to migrate to a more hospitable environment during the change or to become trapped at a natural barrier, such as large bodies of water, prior to the change while they tried to escape. This would have caused us to find the fossilized remains at or near these barriers, not scattered helter skelter about. Plus, these remains would have needed to be entombed in one of only two ways I know of that would deter scavengers and oxidation and still cause the dinosaurs to be caught by surprise. The available evidence on the subject does not support the current line of thinking. A line of thinking that does not even address why the animal remained on location despite its natural desire to survive. We know the dinosaurs were there – without understanding why.

Aquatic animals fossilized in mountain ranges. This is a hard to explain phenomena. We need to develop a theory where either the animals were thrown to their present location or that provides for the total resurfacing of the planet. Either or both of these possibilities could have happened. As a matter of fact, I believe them both to be true. I address this in this book because I have never found a satisfactory scientific explanation in the current dogma for how these events could happen. It is one thing for scientists to tell us that the mountain ranges used to be sea beds. It is another to describe the process steps that would enable this to happen. I won't just accept the snippet of information that science provides, when ignoring of the "why" has such implications. Science needs to address this question from the point of view that an engineer would take. Cause and effect. The cause will be dramatic!

Wooly Mammoths in Permafrost. Like the dinosaurs written about above, we find an animal that supposedly remained in a hostile environment despite their survival instinct. But, that is not the most bizarre thing in this anomaly. That belongs to the freshness of their flesh today. These animals could be thawed out and eaten right now. The meat has not spoiled and it has not suffered from freezer burn.

We know that frozen meat cannot be eaten and it will keep without spoilage as long as it stays that way. The freezers in our kitchens demonstrate that for us every day. So, if these animals stayed in that state for thousands of years, we are bound to find them that way now - safe from scavengers and predators.

However, the wooly mammoth, like our present day elephant, were a warm weather species. Yes, they have what appears to be a fur coat; but, this hair probably was better suited for cooling. This is evidenced by their other body parts that were adapted for hot weather. Those large ears are one of those parts. Ears that, if exposed to extreme low temperatures, would have suffered from frostbite.

Another major oddity of some of these animal findings is the status of the food content in both their mouths and stomachs. Food that is not likely to have flourished in these locations, let alone remain in the mouth not swallowed and, if swallowed, not digested. I would think that just before I die that I would either swallow or spit out anything in my mouth.

And, as to the stomach content, I wonder what halted the stomach acids from continuing to breakdown some of the food. I understand that the digestive process eventually stops at death; but, does it happen so fast that some of the food is virtually unaffected?

It would seem that the process that put the wooly mammoths in this state would have to happen extremely quickly. Flash freezing that prevents a swallowing reaction, stops digestions rapidly, and saves the flesh from suffering from frostbite. I have not heard of any scientists looking into this phenomenon. So, how do we explain this anomaly?

<u>Flood Theory Support</u>. Aquatic animal fossils in the mountains, strata layers, unexplained water erosion marks, and a great number of other items that you may know about suggest that during periods of times in Earth's history there have been great floods. While myths, legends, and religious writings have told stories on the subject; I am not looking at these apparent witness accounts. This is because we cannot cross-examine the

tellers of these tales. The courtroom shows we have seen on T.V. have shown us that this would be considered hearsay. It would be ruled as inadmissible by the judge. So, what are we left with as evidence? I guess we call it forensics - hard evidence.

The items you have seen above are just the initial exhibits. You are on the jury and more evidence is coming. Please examine everything before you formulate your judgment.

Chapter Two

The Creation of Fossils

Defeating Erosion

The creation of fossils is a question that is often ignored by the sciences. Finding them and applying an age to them is common. Yet, we fail to ask how was this made, especially, when the fossil is on the surface of this planet like the ancient foot prints found around rivers and streams. How can a foot print in the sand or muck withstand the flow of water?

We have all built sand castles and without noticing formed foot prints in the sand at beaches all around this planet. These structures and impressions have always been temporary. Even when the waves and tides don't erase our marks on the earth; rain does, not to mention the effects of wind. None of us has left an everlasting signature. Erosion works!!

When it comes to bones and other organic material, placement in an oxygen free environment is required in order to leave an eternal impression. Now we have to find the right conditions that would facilitate fossilization.

The remains of an animal killed by a moving automobile will be picked apart by scavengers and any leftovers will rot and decompose into nothingness. Even bodies placed in caskets and vaults after embalming will decompose and return to mother earth. Eventually, even the vault and the casket will oxidize. The mummies that have been found are not examples of defeating oxidation – they are only proof that the process can be slowed down. This is proven by the fragile state of these mummies now located in museums all over the world. Upon exposure to the surface of the planet, they become susceptible to the very elements that those who performed the mummification were trying to avoid.

Even if mummification could stop oxidation, we would be even more perplexed by the thought that someone or something would have had to perform this process on thousands of bones that have been uncovered. I doubt that there was a society that disemboweled and/or embalmed dinosaurs, so that they would leave a legacy that lasted eons.

Other organics, such as leaves, grasses, and woods would also not survive ages to be found today. Decomposition happens. Today, we create compositing contraptions in our yards to deal with this type of debris. Nothing magical occurs in a compost pile. Many call it – rotting. Basically the material is oxidizing. As it breaks down, it turns into a material similar to dirt that we use in gardens and flower beds to feed nutrients to our beloved plants.

Just add oxygen to these organic materials and oxidation (rot, decomposition) will take place. That is why we periodically turn the material over with a shovel. Which means - if oxygen is present, the process will take place. Can anyone think of and defend a way in nature that oxygen can be totally removed from anything?

The sea, mud, and muck also do not totally remove oxygen from those items that will become fossils. After all, muck is the process in action. Materials gathered together and decomposing are the ingredients of muck.

Proof of this oxidation in a low oxygen environment is exampled by the USS Arizona in Honolulu Harbor. This memorial submerged in the water is oxidizing. It is slowly wasting away. While it is at a slower rate than it would be if it were above the water, it is still succumbing to the process. Mud and muck would probably make the change slow to a crawl; however, it would not stop.

So, how can the fact that the fossils exist be explained? What process in nature could happen that would preserve any organic material long enough for debris to build up on top of it until the weight creates enough pressure to form a fossil?

One answer is volcanic ash cover. A complete burying of materials under this ash could explain the number of fossils found. Yet this causes one to wonder about the conditions required that would create global fossilization within the same strata layer. What would cause volcanoes to erupt all around the earth at the same time?

Another answer of the same type could be the covering of the bones by debris returning to the earth following a meteor impact. This would adequately accomplish the protection from oxygen. How many times has this event happened to the extent that would cover the entire globe and create the innumerable strata layers?

The only other answer can be found in any modern kitchen. It is simple - we rely on it most every day: Freezing. In our freezers, we find numerous examples of items that are defying oxidation. As long as a roast is frozen, it will remain and exist. The same would happen to any organic matter placed in the freezer. While freezer-burn may cause us to throw the roast away, it won't allow the oxidation process to continue. If it's frozen, organic matter will not oxidize. Plus, scavengers will not eat frozen meat.

If some organic item - say a dinosaur - becomes frozen and drops in its tracks, it will remain untouched by scavengers and defy oxidation. Subject to a continual frozen state - this same dinosaur will eventually be covered by all manner of materials through erosion deposits, space/sky dust accumulation, lava and the like. Over many years of this build up, the weight will increase. As the pressure builds from this weight, the process of fossilization begins.

Granted, as the pressure goes up, so will the temperature; however, by then the oxidation process has been slowed so much that fossilization takes place before the dinosaur bones rot away. Thus, a fossil exists for a future generations to find.

A flash freeze fossil creation method is really the only one that I think could protect foot prints in sand long enough to be made into fossils. The other methods would actually fill the prints and subsequently prevent their being found on the surface of the earth.

The explanation of how this freezing would take place will be addressed in another chapter. The global covering by volcanic ash will become obvious later, when events that would cause the vast amounts of lava to reach the surface will be discussed. The meteor impact theory is not covered herein because the event probably happened a few times throughout earth's history; but, I doubt it happened once for every strata layer we have.

Fossil creation is an important piece of the puzzle. Their existence begs to be explained. The explanations cause greater questions. The questions lead to theories. The theories point to cataclysms. All of this warrants our attention.

Chapter Three

The Grand Canyon

Creation by Tectonic Movement

A friend of mine once asked me how I thought the Grand Canyon was formed. We discussed the possibility that river erosion as the accepted cause was the reason. We both disagreed with that dogma.

He had a problem with the idea of the amount of flowing water that would have been required when you look at the size of the gap. My problem with river erosion was not the amount of water; but, its velocity that would have had to be maintained. Both arguments were valid.

However, an alternative for the canyon's formation was missing. I now believe I have formulated a theory of my own.

My thoughts are easily verifiable given the resources. These resources are beyond my economic means. It will require grants and donations that my qualifications do not command. Hopefully, someone will take on the task and check my idea.

Before we talk about my theory, let's look at the problems that seem to me to be apparent with regard to river erosion creating the Grand Canyon.

First, my friend was right. The amount of water to cut the canyon would be incalculable because we would not know the average depth of the river. Let's suffice it to say that it is "a lot". So, where did this water come from?

Even glacial melt would have its limits. The melt would only be able to provide a set amount of water. Then, through some sort of repetitive process, all of the water would have to be concentrated on one path. Water following fluid dynamics will travel the path of least resistance. So, the trail that this river followed would have to be the lowest elevation possible on average on the way to the sea. This is not true today. There are spots that, when connected by fluid resistance measurements, would have given the water from the melt an easier path to the ocean. A number of existing rivers are proof of that.

Now that other rivers have been mentioned, how come they have not also created canyons on that scale?

My argument against river erosion dealt with velocity. The speed at which the water would have to travel on its entire trip to the sea would have been very fast. Note: I said the entire trip. Anywhere that the water slowed down, it would have released some of the debris that it was dragging along. As this debris settled to the bottom, it would start to create a dam. To overcome the damming effect, the water would need to find another path or build a lake, that it eventually would have to spill over. This spill over would be a widening similar to the mouth of any other river. Proof of ancient lakes and/or the actual bodies of water do not exist. So, the high velocity of the water would have had to been continual without pausing at some sort of damming structure.

As proof of what I am saying, I offer that the Mississippi River needs to be continually dredged in order to keep it open to shipping traffic. Without human intervention; Ole Miss would widen, shallow out, and thus, slow to a crawl. The more it slows, the more likely a lake would form. These types of regions are called “deltas’ to describe the effect. The Nile and Mekong Rivers are other examples.

Now let’s look at both of these arguments against river erosion together. The large amount of water in a slow (normal flow) river would be greatly compounded. A pipe carrying water at one foot per minute may transport 10 gallons; yet, that same pipe moving water at 3 feet per minute will carry 45 to 50 gallons. The amount of water to defeat the combined arguments becomes insurmountable in my mind. Where did it come from and where is it now?

So, what process could create the Grand Canyon?

Let's look to plate tectonics and the movement of them. The earth's crust is floating on magma in large pieces. A virtual "party-barge" similar to large numbers of boats tied together on a lake, drifting on liquid. Each piece is referred to as a plate. We most commonly see them as land masses, even though many continents and ocean/sea regions are made up of multiple plates.

Where two plates push upwards against each other, mountain ranges are formed. With that in mind, the reverse process would happen if two plates pulled away from each other. You would now have a canyon on a grand scale. I propose that is what has happened and is continuing to happen.

I further believe that the reason we do not see an opening with molten lava showing is that these plates, relatively freely moving, have a subterranean plate under them. This plate now has a river of normal size running generally north to south on top of it.

So, now the strata layers exposed on the walls of the canyon with sharp, jagged edges would not need vast amounts of water and erosion to explain the canyon's existence. The lack of water marks on the canyon walls would be self-evident. In short, in that region, the land mass is growing.

This theory is easily tested using seismographs and GPS devices. By positioning them at various spots around the canyon's rim, we could monitor for movement. We would answer the question of how the Grand Canyon was formed. These tests might also be merited in the area where the Alaska/Russia land-bridge was supposedly connected and on Iceland where there is the appearance of a mini canyon forming.

The reason for the inclusion of this topic will become apparent when you read Continental Drift, Pangaea in Reverse.

Chapter Four

Continental Drift

Pangaea in Reverse

Having discussed the formation of the Grand Canyon by the movement of tectonic plates; the question of continental drift comes to mind. If the land mass of North America is apparently growing, then is the earth growing?

My answer, of course, is yes! Due to the accumulation of space dust and objects that are pulled in by our gravity, this planet is growing. However, that is not the growth that I am questioning. I want to know if earth is expanding from the inside out, growing outwardly, if you will. I propose that it is not.

My theory is that as some areas are expanding, others are contracting. This expansion and contraction is in a relatively equal rate. The surface area of this planet remains approximately the same.

The Pacific Rim, the raised elevation area around the ocean, is a hot bed of tectonic and volcanic activity. The tectonic movement can be a push or pull movement or a side to side slide against each of the plates; but, a volcano is an obvious expansion. There is little doubt that where there is a volcano - there is an expansion of the earth.

The molten lava is pushing its way to the surface. As the lava forces itself against the surrounding area, room has to be made for it. The entire surrounding area fights to hold its position in a losing battle. The force of the molten lava, as it rises to the surface, has to be pushing the things in its way to the side, if you will. If the matter that had been blocking the path of the lava did not give way, the volcano would not exist. So, where there is a volcano - there is expansion.

The expansion has been discussed as a lead in for contraction. Unless this planet is growing at a huge rate, you have to have some points of contraction. The sciences call it seduction. An example of a point of contraction lies just off the west coast of North America.

Two tectonic plates are exerting their force on each other. Just like two sumo wrestlers, these plates are fighting for the same position. Both trying to stand their ground. And, like in sumo wrestling, one eventually gives ground. One plate ends up under the other. We know the shock of this event as an earthquake and its aftershocks. But the process as it relates to the tectonic plate is not over.

Once one plate succumbs to another and ends up pinned beneath the weight of its opponent, it begins to return to a molten state. It becomes trapped in an environment of extreme temperature and, more importantly, extreme pressure. This pressure is so intense that heat rises to almost unimaginable temperatures that melt everything.

Thus, the plate that ends up under the other does not remain there. It actually returns to a state of molten lava. It joins the liquid layer of this planet awaiting its return to the surface in the form of a volcano. A Process of Perpetual Renewal. So, how do we determine where volcanoes are most likely to erupt?

The answer is simple. Where the path of least resistance is, you will find the locations of expansion. Where earth's crust is thinnest. If the crust is thin, it offers the path of least resistance. This is stated to provide a lead in for identifying the points of contraction. This is a problem that is also easily addressed.

Points of most resistance are the most likely

contraction points. Mountain ranges stand out as candidates. As the plates push against each other, they force each other into the air - rising to allow for the areas of expansion.

However, there is another point of contraction to look at. The lower elevations - below sea level. This is where two plates can come in contact with each other and push equally downward. Both plates are succumbing. Neither plate has won the day and remained above the other. Pushing at one another in a "V" pattern.

In this case, both plates are returning to a molten lava state. Completing the cycle in a relatively accelerated rate at the intersection of the "V". Making room for expansion at other points and causing the surrounding high ground to move towards the center of the "V" shape. With this type of seduction in progress, we will find two coastal lines moving towards each other.

It must also be pointed out that the rising of the

Atlantic Ridge, a mountain range forming in the water, contributes greatly to the seduction.

This is what I am proposing is happening in the Atlantic Ocean. The American, European, and African Continents are not moving away from each other as proposed in the Pangaea Theory. In the Pangaea Theory, scientists believe that three large continents were, at one time, connected to each other. Then millions of years ago they separated. Basically, they floated away to their current positions. I think they are doing the opposite. When you look at the apparent growth in the Pacific Rim and the creation of the Grand Canyon, you have to find a region that is succumbing and is acting as a contraction area. The Atlantic is a good possibility.

Look at the coast lines of the continents involved and you will find clues. All of these continents have what appears to be drop offs. Sharp, hard coast lines; that do not gradually slope to the ocean. As a matter of fact, the west coast of Africa has exposed strata layers as if the land that used to cover the sides has collapsed into the sea.

Current dogma says that this exposure is caused by one tectonic plate pushing over the top of another. This seems unlikely. Remember the sumo wrestler example. One plate would have to be pushed down; thus creating a seduction point. And when two plates push equally, they will form a mountain range. But the African coast does not fit either of these scenarios. It truly looks as if a form of erosion is at work.

All through the Atlantic coastal regions you find these examples of erosion. Iceland, Greenland, England, Europe, and, of course, North America appear to be dropping their coastal high ground into the sea. Erosion at work!

We even have an example of this type of land collapse in the news and on documentaries all of the time. This is the Island of La Palma. Geologists, volcanologists, and many governments are closely monitoring this Island. Watching for the western coast of this isle to fall into the ocean. They are paying a lot of attention to it fearing the tsunamis (tidal wave) that it will create. They are not looking at the indication that the landslide will be a result of a distant seduction.

The best examples that land masses are moving together in the Atlantic region are the caves below the Island of Cozumel, Mexico and the movement of the lighthouse located at Cape Hatteras.

Below the surface of Cozumel, there are a series of caves that are below sea level. The water within the caves is a mixture of both fresh water traveling to the sea in the form of an underground river and saltwater that has forced its way into the caves because the tunnels are below sea level.

These caves drew the attention of archeologists, when

in the 1990's, a group of divers found Mayan artifacts and bones deep inside the tunnels. Expeditions were launched to explore the caves to find an explanation for manmade items and bones being located nearly a mile inside the tunnels. The exploration crew was able to find a plausible explanation for the dilemma when it was dry. After all, it was very unlikely that the Mayans were able to dive to the location of the artifacts.

This is not the point I want to address. The item of importance is the fact that one time the caves were not under water. You are probably wondering why I said the "fact". Well, the answer lies in the formation of stalactites and stalagmites. This underwater series of caves contain these geological features. Stalactites hanging from the ceiling and stalagmites rising up from the floor. So what - you would expect this in a cave - right?

Yes - if the cave was not underwater. Stalactites and stalagmites are formed by water that is dripping down through the rocks and carrying material with it. As the water dries, it leaves the rock material it was carrying behind. Over a very long time, this material builds up to create the formation. Thus, because the water that is dripping needs to dry (evaporate), it proves that the cave was at one time dry. Otherwise, the surrounding water would have washed the sediment downstream and prevent the formation of the columns.

Therefore, if the caves were at one time dry, then they had to have been above sea level. To have been above sea level they would need to be at least 50 feet higher in elevation than they are now. The island is sinking into the sea! With my theory of "Pangaea in Reverse", this would be expected. I cannot think of another way to explain this drop in elevation.

I would like to discuss the artifacts that were found in the caves for a moment in order to explore a timeframe for the sinking. Let's assume that the Mayans were walking in a dry cave nearly a mile to deposit the items that were found. This would give us a rough estimate of when the cave was dry. As early as 1000 years ago, this cave could have been dry. This would give us a rate of drop in elevation of up to a half inch per year.

Now let's look at the lighthouse at Cape Hatteras, North Carolina. Here we find a structure that has been moved in order to avoid falling into the ocean. The lighthouse was built in 1870ce at a distance of 1500 feet from the water's edge. In 1987ce, measurements were taken that found the lighthouse was then only 150 feet from the water's edge. A dilemma that was corrected by the subsequent moving of the lighthouse. It sure looks as if the shoreline has been washing away. A rate of erosion in the neighborhood of 11 feet per year. It would have been nice to know the original elevation of the lighthouse; but, that information either does not exist or is too hard to locate. With that information we would have been able to calculate its rate of sink.

If the east coast of the Americas and the west coasts of Europe and Africa are falling into the sea at these rates, we must be able to measure it. This is a significant drop and slide in comparison to the rise of the Himalayas. If we can measure the growth of a mountain range to that small amount, then we can measure an equal amount in drop and a slide displayed by Cape Hatteras. It should be almost visible. Someone should get busy looking into this. Maybe existing GPS data on International Airports can confirm this.

Knowing that we have a sinking of land masses in the Atlantic region lends credence to my theory that the movement towards the middle of the ocean is valid. It would also cause one to think of the shape of the coast lines around the Atlantic that appear to fit together like a puzzle. I envision the shaping of the coast being caused by the land masses following the contour of the mid-

Atlantic fault line.

So, if I am right; where is all of the material going?

I believe that the trench that runs generally down the middle of the Atlantic Ocean is actually a point of contraction (seduction). There are two sets of tectonic plates that are pushing into each other and both sets are succumbing. The "V" pattern mentioned earlier, in grand fashion. Both sets quietly returning to a molten state and completing the expansion and contraction cycle.

If this is true, then Pangaea has not happened yet! The super continent that scientists talk about is not a thing of the past; but, a possible outcome for the future. This, like the theory of the creation of the Grand Canyon, is a hypothesis that is easy to verify. If the regions are moving towards each other we should be able to measure it. After all, there are people monitoring the growth of the Himalayas in Asia. Hmmm - growth in the Pacific Rim.

You must be, at this point, wondering why I am writing about this. You may think I am just trying to appear smarter that scientists. Not so! I am trying to show that the earth's crust is on the move. That it is not stationary. It can be changed in its position as we know it. After all, it is floating. So movement should not be too far from our ability to comprehend. It is similar to blowing on the filmy substance that forms on hot chocolate when it cools. The film jiggles, vibrates, and moves, if enough force is applied.

So, why wouldn't the crust of this planet?

Chapter Five

Creeping Salt

The Formation of Salt Flats

This chapter can be interactive, if you wish. There is a little experiment discussed that can be put together to explain to yourself how a salt flat is formed. The materials needed are simple and can be found in the kitchen. I urge you to try it.

Take a pan filled with water and bring it to a boil. At that point, begin to add salt until total saturation is accomplished. This will be evident by the salt particles that are left at the bottom of the pan. When no additional salt will dissolve, the water will be holding all the salt it can.

Then put some of the mixture into two bowls. There need not be anything special about the bowls. They can be standard, the type a child uses to eat cereal.

One of the bowls can be left exposed on the kitchen counter or anywhere that it will be undisturbed at a temperature near 70 degrees Fahrenheit. The other, if the season is correct – winter, can be placed on a table out on your deck or patio, also exposed. An off-season alternative would be a space in the freezer. The experiment is now set into motion.

Nothing immediately happens. Days will elapse before anything interesting develops. So, this is an experiment that you set and forget for awhile. Easy!

Enough about the experiment's setup. Now I will tell you about the results that you will observe.

The bowl that remains on the kitchen counter suffers

from evaporation to the point of complete removal of the water. Nothing but salt remains. The bowl that is outside or in your freezer shows little effect of evaporation. However, it will still be filled with a saltwater mixture that will not succumb to freezing. It will still be in liquid form. That is not unexpected because saltwater does not freeze. The water needs to release the salt before it will turn solid.

What will be interesting is what will happen to the salt in the bowls. The bowl that is indoors and had totally evaporated shows signs of what I will call "Salt Creep". It appears that the salt molecules have tried to hitch-hike with the water vapor as it rises out of the bowl.

A crust layer will climb up and over the side of the bowl. Yes, the salt will have risen above the original water level and if you think to look for it, there may be salt on the counter-top around the bowl. Even more amazing will be the bottom of the bowl or more specifically, the remnants of salt at the bottom. The salt will not settle into a salt flat. In fact, the salt deposits left will be in the form of ever shrinking circles. This shrinking continues until the very middle is what appears to be a tiny, salt-walled cup that has barely a trace of salt in the center. Not a flat at all!

The saltwater filled bowl that is left exposed to the elements of the outside's winter weather or in your freezer will show a greatly different result. Because there has been some evaporation, you can assume that some salt creep has taken place. However, the evidence of such will not be visually apparent. What will be apparent is the layer of salt that is at the bottom of the bowl. A salt flat will have formed. A very even surface of salt will be at the bottom with no evidence of a cupping form. It will be as flat as those found around the earth. Yes, there will still be water over it; but, a flat will have been created. Additional time or a breaking of the bowl's side would remove the remaining water.

This creation will be caused when the water temperature drops and its saturation point changes. The water's ability to keep salt in suspension will be reduced. Thus, the salt will return to its solid state and will sink to the bottom.

The surface of the solid salt will become even in the same way that dirt settles in a pail of water that is used to clean floors. The heavier substance sinks. There will be no hills and valleys. There will be no ripples or any other defects. Like I said, a salt flat will be formed.

This experiment should confirm for you that evaporation of trapped seawater did not create our salt flats. If that process had been the cause of the phenomena, then there would be salt deposits up and over the mountains that surround the salt flats due to salt creep. These types of deposits are simply not there.

The argument could be made that the rain washed the

salt back into the low area. I offer that rain never erodes just one substance leaving all others. If erosion helped, then the flats would be made of salt and mud, throughout. Also, remember that the deposits are flat. The total evaporation process would have left a cupping effect as the water was removed and this effect would be compounded if rain washed salt back down the mountain and became vapor prior to reaching the center of the cupping form. Which would have happened due to the relatively small amounts of water that could have been present over such a vast area.

So, it becomes evident that in order for a salt flat to form, extreme low temperature need be present in the process. Thus, wherever there is a salt flat, there was once an exposure of saltwater to a freezing condition. The process to explain this will be addressed in another chapter.

Chapter Six

Oceanic Conveyor Belt

Earth's Life Blood

Have you ever wondered why great portions of North America are a tolerable temperature and areas of Siberia at the same latitude are frozen?

Do you also ponder why England and the Norse regions of Europe enjoy moderate climates despite their northern position on the earth?

As you may have figured out already, this phenomena also applies to Japan and other Asian countries and affects the southern hemisphere similarly. It is the result of Oceanic Conveyor Belts.

The Oceanic Conveyor Belts are the flow of water that occurs as a result of the natural heating and cooling of the water. These belts run from the equatorial regions and transport warm water to the polar caps, both north and south.

While these belts deliver warm water to the colder portions of the earth, they bring with it warm air and weather. These warm air and subsequent weather patterns are the result of thermo transfer. It is the life blood of the earth in these temperate high latitude regions. Without this heat transfer, these areas freeze.

Evidence of this process is in the comparison of the northern European regions and the Siberian region. Siberia lacks the benefits of the Oceanic Conveyor Belt. This region is too far away from the sea to enjoy the warmth delivered by the process. A process that works on a simple principle of convection within saltwater. As a result, this region in its northern latitude is in a deep freeze while the northern European regions in close proximity to the ocean are comparatively temperate.

Supporting evidence of this process is provided by the much higher average temperatures that are present in the southern tips of Africa and South America, where both have two Oceanic Conveyors working nearby, despite their close proximity to the southern polar cap.

As an example of how it works, let's look at the North Atlantic Ocean. Saltwater at the equator is heated by the direct exposure to the sun's rays. It naturally moves in the direction of the colder water, just like the heat moves in a convection oven in your kitchen. As the warm saltwater moves north, the water there already needs to make room and get out of the way.

The process that enables the making of space for the warm saltwater is caused by the fact that the salt water's density goes up as the temperature drops. Thus, it is heavier and sinks. Cold water can not hold as much salt in suspension as hot water, thus as the temperature drops it becomes heavier by volume. The temperature drop is caused by the thermo transfer discussed earlier. The water warms the regions around it and brings a climate that is much more tolerable to live in.

Of course, as the saltwater becomes heavier, it sinks. As it sinks, it makes room for new warm saltwater to take its place. As this process continues, the beginning of the conveyor is in place. Now, we have very dense, cold saltwater on the bottom of the ocean that is being pushed back to the equator. Now what?

Well, if nothing changes the density and temperature of this saltwater, a blockade forms. The Oceanic Conveyor Belt stops. All of this dense, cold saltwater would pile up on itself. No more movement unless something intervenes.

This intervention comes in the form of mother earth.

More specifically, the heat from molten lava in the shallow portions of the earth's crust and heat vents. As the saltwater moves along the bottom of the ocean, it comes in close proximity to earth's natural heat due to the thinner crust and the vents which produce water temperatures that exceed 650 degrees Fahrenheit in the immediate area. This saltwater also gains energy from the immense pressure from the vast amounts of water pushing down from above. This super heated water mixes with other cold water and achieves warmer temperatures. This is why the water does not freeze on the bottom of the ocean despite the lack of light. As the saltwater heats up it rises back to the surface and is again reheated by the sun. Which enables it continue with the process steps. The cycle is complete and the Oceanic Conveyor runs.

The Oceanic Conveyor Belt performs its purpose using a delicate balance of heating and cooling saltwater at a pace that has been in place for eons. Continually stepping through the, without missing a beat. Well, Almost!

The process requires a set amount of salt in the water. If this ratio is correct, the conveyor belt runs at its optimum performance. Our global environment is perfectly balanced for continued existence of life forms. If the mixture changes, so does the belt's operating speed. Increase the salt to water ratio and the operating speed of the conveyor belt increases. The faster it runs, the warmer the surrounding land regions become and the cooler the equatorial temperatures are. Up until too much salt is present and then the polar caps melt away and again, the shutdown of the Oceanic Conveyor Belt occurs. This is due to a desalination caused by the introduction of huge amounts of fresh water.

So, we can imagine an upper limit in the salt to water ratio. I don't know what it is; but, many a scientist could calculate the ratio, if we need to. However, the upper limit is not our immediate problem. The opposite is the crisis at hand.

News reports have been reporting that our seas are desalinating. The ratio of salt to water is decreasing. I will explain why later. For now, I will talk about the expected results, if the desalination continues.

If increasing the salt content will speed up the Oceanic Conveyor Belt, then the opposite should happen if the amount of salt decreases. As the ratio of salt to water drops, so does the operating speed of the belt. A slower belt will eventually result in colder northern/southern regions. Regions we know as temperate will become bitterly cold. This is because the thermo transfer takes place at more equatorial regions and there is not enough energy remaining to warm the other areas.

However, the first indications that the belt has slowed down is a dramatic melting of the polar caps caused by the warm saltwater remaining in close contact with the ice longer. This is a problem that we face now. It is a change in the cyclic process that gives the impression of the opposite effect. The saltwater that is passing by the caps still has energy to transfer to the caps and because it remains in the area longer, it passes the heat. This extended stay obviously results in melting ice. Thus, the saltwater that sinks at the polar caps has dropped to lower than normal temperatures as it makes room for the arrival of new saltwater.

As the operating speed drops further, the thermo transfer will be complete in regions closer to the equator. This future lack of warm saltwater passing by will finally result in a decrease in polar temperatures. The saltwater that does pass by will have cooled in the regions closer to the equator and will have given up all the heat (energy) it could. If all of the thermo energy has been absorbed and new warm saltwater is not forthcoming, the regions will eventually succumb to the otherwise natural state of permafrost. Just like Siberia, with nothing to gain energy from, the region freezes. An apparent ice age is beginning.

This is just one element of the disaster that is in our future. The other is that because the equatorial saltwater is not making the trip to the polar caps to cool down, it heats up and through thermo transfer heats the regions where it is. The equatorial regions would then begin to cook and depending on how slow the Oceanic Conveyor Belt is running may even boil.

We end up with two totally out of control climatological events happening at the same time. Both of these will be at the opposite side of the thermo spectrum. One heating to extreme high temperatures and the other cooling to unimaginable lower relative temperatures. Trouble for civilization will have been set into motion due to desalination.

The reason for this desalination is covered in another chapter. Keep reading - we face trouble; but, we are not powerless.

Chapter Seven

Ice Ages

Events That Didn't Happen

Ice ages are the theory provided by science to explain why there is evidence of glacier ice having covered areas around the globe that are not in polar regions. Striations and depressions left in the earth give the impression that at one time ice had spread over landmasses far outside the polar caps. An ice age provides a quick answer. An answer that fails to address all of the evidence available and that does not provide for its cause. For every effect there has to be a cause. Ice ages do not provide a plausible explanation for what caused them. Therefore, we need to look at the theory's plausibility and determine if they can be created in the first place.

Just because it appears that an ice age happened does not make it so. The theory of an ice age has the answer to some of the geological formations that scientists have found, but not all. It would explain striations carved in the earth by ice flow; if the trail of marks led to the current polar caps. They don't. Depending upon which marks you follow, you trace them back to various spots on the planet that are not at the caps, as we know them. The destinations that these marks lead to are in completely different locations. This gives the indication that the polar caps have been located elsewhere. Further, where the striations intersect you find depressions in the earth's crust. This would be expected when you consider the immense pressure that the ice would have put on the crust.

Ice ages don't even come close to offering a cause for strata layers. While it is true that an environmental change would have taken place; it is also fact, that the change would have been gradual. The climatological change would have happened over a relatively long period of time. The ice would have overtaken an area slowly. This slow change would appear today as a blending of the strata layers. A definitive, hard line strata layer would not be present.

Even more oddly, is once the ice moved back, if you will, at the end of the ice age; why did not the original environment return?

If the area was previously a forest - it should return to being a forest. The seeds of the old forest would sprout trees after the ice retreated. The position of the ground would not have moved from its original position with regard to the polar caps and equator. Therefore, the environment would be exactly the same as it was before the ice age.

Without a crustal shift, the forest would become a forest again following an ice age. From one ice age to the next, you would find forest in the interim. Forest, ice age, forest, ice age, and so on and on - forever.

Also, if ice ages had happened, why did the survival instinct of animals quit working?

Try to imagine a wooly mammoth or a mastodon, both warm climate animals, ignoring the ice approaching. Defying their survival instinct, these animals remained in place. No movement to more temperate areas. Staying to the death. No way!!

Further, imagine the first Americans, who supposedly migrated across the Alaska/Russia Land Bridge. A hunter/gatherer would not follow ice as it withdrawals. His life preserving instinct would cause him to move away from the ice towards happier hunting grounds. Following the sun would be the appropriate path to livelihood. As he traveled, he would learn that the higher the sun is in the sky, the warmer the weather and the more hospitable the living conditions. This nomadic life is one of movement in order to find the needs for survival.

I am not arguing that man didn't cross the land bridge. I am orating that the position of the bridge was different. I believe that man would have crossed it, if the path towards the sun followed it. The only way for the path over the land/bridge to follow the sun is for it to have been in a different spot relative to the poles than it is now.

Position a globe so that the Alaska/Russia Land Bridge fell in a more east/west line and then you would see a reason for its crossing. In a search for fertile ground for crops and abundant prey, a primitive man would move horizontally from place to place. This way he maintains a relatively similar climate along his travels. He does not have to change his style of living and learn new things to continue his survival.

If you have positioned a globe as suggested, you can now picture what the earth may have looked like in a previous epoch. You may even understand that the land/bridge could have been apparent, when you imagine a massive polar cap holding water in a frozen state. Lower sea levels as a result of larger than normal caps.

Now look at this globe and see if you can see evidence of previous polar caps in your new north and south polar regions. Depressions and striation marks will give away the positions. You will then notice that those regions will now give us the appearance of having suffered an ice age. No explanation for environmental change is needed to give a reason for the evidence of ice. It will become obvious why there is evidence of an ice age in various parts of this planet.

Still need more evidence that ice ages did not happen the way we were taught?

How about trying to find an explanation, in the old ice age theory, that answers the question that rises when core samples from the ice covering Greenland found vegetation underneath. Yes, identifiable plant life was/is at the bottom of all that ice. Thus, at one time, the ice was not there. A polar region with no ice on it seems very unlikely. Sure, if the surface was super hot, there would be no ice. So, it happened millions and millions of years ago. I don't think this was possible. Too hot for ice in that region means it was probably too hot for plant life. Also, the vegetation is identifiable. What happened to the evolution of the plant life?

The plant life remains the same while everything else evolved - Strange!!

In closing this chapter - I ask "What process could take place that would cause an ice age?"

My theory of the Oceanic Conveyor Belt's shut down is a trigger for a different event - a crustal shift which will provide a viable answer for ice age evidence. It provides for cause and effect. It will identify the subsequent events that will leave the apparent proof of an ice age. It will not require unmentioned miracles to happen for the ice to have been in temperate locations. The current explanations for ice ages would need the defiance of physical laws for their creation. Furthermore, man would not be the cause of it.

A meteor impact theory is weak because gravity works. Any debris that goes up must come down. It may stay in the atmosphere for a long time; but, not for thousands of years. This extended period would be needed to block the sun long enough for all of that ice to grow. That isn't the weakest part of the argument. That weak link belongs to the fact that in this scenario everything would freeze in place. Debris blocking the sun's rays, would block the rays everywhere. This would shut down all weather. With no heat to generate vapor means no precipitation. With no precipitation, you have no new ice because there is no snow.

This problem of no precipitation is also part of the reason that none of the other ice age theories work. As an example, if a great number of volcanoes erupt at the same time and blocked the sun everywhere, it still stops the making of new snow. The only other theories for ice ages I know of are: The Ice Ball Earth scenario, in which everything froze into a giant snowball. This proposed event has no feasible cause. The other is the possible moving of the planet in relation to its orbit around the sun. Supposedly, the earth has moved away from the sun, had its ice ages, and then moved back to a more proper orbit. Sort of like a yo-yo. O.K. if this is true, how did it happen. They seem very farfetched and beyond the limits of the Laws of Physics.

These theories provide for total freezing of the planet; however, these types of events would leave evidence that would look similar to the Siberia region. Permafrost would have been everywhere that was not already covered with ice. No new ice would be made and there would be no glaciers that expand.

These current theories are extremely weak and fail to address the cause and effect principles of any process steps. Remember, that if you don't use all of the available evidence to support a theory and you can't find a cause, then the theory is probably wrong. I don't think I need to say more. What do you, a member of the jury, think?

Chapter Eight

Crustal Shift

The Cyclic Catastrophe

The recurring disaster that we face is simple. It is easy to understand. It is not hard to imagine the result. The problem lies in trying to convince people that it has happened and, more importantly, that it will happen again, unless some sort of intervention takes place.

In a nut shell, the disastrous situation is the sudden, violent shift of the earth's crust on top of the molten lava that lies deep inside the earth. Remember that the crust is merely floating on the mantle. I illustrated this in chapters three and four. The tectonic plates are virtually boats tied tightly together; at sea in an ocean of liquid matter (magma). We have all been told this. The movement of the tectonic plates are discussed on the evening news all of the time following an earthquake. The crust is moving - nothing new.

What will be new to many of you is the fact that it can be a massive movement. Yes - I said “the fact”. The evidence is all around us. It can be found on any continent, so you can go look for yourself. If you review books, websites, and articles, you will find proof. And if that is not enough for you, just dig a hole in your yard. You’ll find the most convincing piece of evidence. One that modern science has given a weak reason for its cause and then ignored it.

This piece of evidence is called strata layers. The formation of a strata layer is caused by a climatological change. The sciences confirm this. OK - a climate change enables the creation of a layer. Deserts turn to water then to stone and any number of different changes. From one to another to another and so on. Over the ages this has happened. My question is "how?"

What could happen to cause the climate in any given area to change?

What could happen globally to change the climate in all of the areas?

What could happen to cause watery areas to become dry and vice versa?

What could happen to make these changes appear suddenly, as if overnight?

The answer I propose is not my original idea. It has just been refined by me. I have narrowed the shift to just the crust like Charles Hapgood had. Not a total earth roll over that Brown had proposed. I found more supporting evidence that did not require the belief in a previous advanced civilization. I even think that I have identified the cause, which has been addressed in another chapter.

So, for now, let's look at the extremely possible crustal shift. This theory is now being used by some Italian Geologists and Archeologists to explain the formation of strata layers and to offer a plausible reason for environmental changes that are evident at their dig sites. They too, see that a sudden change would be required to form a strata layer.

A sudden change is required because a slow change in the environment would have caused a blending of strata layers. No definite line of separation would be apparent. In a gradual change scenario, the layers would have changed slowly. The lower layer could be sand and the upper something else; but, where they meet would be a mixture of both. The separation would be almost indiscernible. This blending of layers from a slow change would be even harder to identify when the multitudes of layers are looked at together. Five layers down would look very similar to the top layer as a result of blending caused by a gradual environmental change.

Thus, because the layers have a definite separation line, we have strong proof that the environmental change was sudden. So sudden that mixed layers are not produced. A crustal shift would provide this process' parameters. A set of parameters that would affect most of the entire planet.

Need more evidence?

Look at the apparent wandering magnetic pole. At different strata layers, we see iron ore that has set up from liquid form to lock in the direction to north, at the molecular level, for that period of time. As a matter of fact, the layer immediately below the top layer points to a different direction than what is magnetic north now. This direction is comparative as it relates to North America. It is close to our current magnetic north. The directional difference elsewhere is more definitive. Meaning that the angle of difference from our magnetic north to the one immediately before changes based on where you are on the earth.

With the wandering magnetic north in mind, we should look as each layer from many points on the earth to determine where previous north magnetic poles were located. This is easily determined through triangulation. Using a globe and strings, we can find the past poles. Lay out strings on the globe that are aligned with the north directions of the past. As an example: One layer down, all of the strings could intersect at the Hudson Bay. That would show you the point of a former northern polar cap. Which is interesting when you realize that ice covered Greenland would fall inside the perimeter of this past polar cap region.

Once that location is identified, you can look for other evidence of an ice cap. A depression in the earth could be there from the weight of a recent cap. The depression may not have been erased as a result of tectonic plate movement yet. Also, evidence of striations in the ground maybe (are) apparent. Erosion not having enough time to remove the scars caused by the movement of the ice. This evidence of movement will also allow us to narrow down the possible center of the cap by the same use of triangulation. With north having been identified, south becomes easy to find, and, of course, the previous equator will begin to stand out. Just rotate the globe to put the previous north in the north and the old south in the south. A different world indeed will become apparent.

Each layer down, we can repeat this process to find former polar caps. Two layers down will give you a different result than the first. The third - yet another and so on and so forth. You will begin to see a tumbling crust that would give you the environmental changes required to create strata layers.

Even more strangely, you will eventually find that a

previous cap will be located at the points we know as salt flats. This will enlighten you to a previous polar cap that was located over a body of sea (salt) water. Much like our current caps that come in contact with the oceans as we know them now. Remember, the salt flat will be formed where the Oceanic Conveyor Belt reaches its most northern or southern point and dives before returning to its equatorial region.

If you need more evidence, you can again look at the formation of fossils. In order to remove the fossilized matter from the oxidation process, in many cases you needed to freeze it. To freeze it for a great many years in nature, you need to bring it to the polar caps. The caps cannot move fast enough to catch a Wooly Mammoth in the midst of grazing, so again, you have proof of a sudden crustal shift. A shift that will drastically change the land- and sea-scape.

Much has been written on this subject. Even

government agencies have used this as a topic in the 1950's. So, why has it been ignored?

I propose that this is not addressed because they (the powers that be) believe there is no stopping it. They think we have to accept this civilization destroying process and that to acknowledge the theory would create mass hysteria. A nothing-to-live for attitude would prevail. A lawless society would take over. A society that could not be governed because the threat would overwhelm us. To us not being able to stop it, I say "Bull".

I realized that every process has parameters that act as controlling factors. Things that have to take place before the end result can be realized. So, I went in search of the key ingredients and how they work. Once I identified the ultimate key to this process, I was able to formulate a plan to try and prevent it. Read on. We have hope for survival.

Chapter Nine

Desalination of the Sea

The Shift Trigger

Having discussed the crustal shift and the formation of salt flats; now would be a good time to outline the triggering mechanism. The reason for previous catastrophic events. The cause of the eventual repeat of this phenomena and, if we act accordingly, the action that can be taken to possibly prevent the same.

In order for the crust to shift, the planet will need to become unbalanced. This unbalance will require an off-setting weight that will overcome the equatorial bulge that has kept us in our upright position for these many years. This counterweight, if you will, is a massive accumulation of polar ice.

As long as the seas remain saltwater, we are relatively safe. The Oceanic Conveyor Belt runs and protects us from trouble. But, that is not happening. The oceans of the world are diluting. The ratio of salt to water is decreasing. This is a natural process. A process that offers an explanation for Global Warming!

The fact that the oceans are saltwater is the result of rivers transporting salts that they come in contact with through erosion. The rivers do not saturate with enough salt to become saltwater themselves. The salt to water ratio is too low. This salt is brought to the seas and then evaporation takes place removing some of the water and leaving the salt behind.

As the seawater becomes vapor and subsequently falls as rain or snow elsewhere, the water left behind in the oceans increases in its salt concentration. Thus saltwater!

Much of the rain and snow falls over land and accumulates as lakes and ponds. However, a great amount flows down rivers back to the sea, perpetuating the continued flow of new salt to the oceans. Why the oceans have not become totally saturated with salt will be discussed later in this chapter. At this point, suffice it to say that the process has been in a delicate balance for eons. But, it won't last forever!!

Due to a great many years of erosion, the salt deposits that the river water comes in contact with are depleting. Less and less salt is being added to the oceans. That in itself is not the problem. The dilemma comes from the formation of future salt flats.

The saltwater that is enabling the continued run of the Oceanic Conveyor Belt at its proper operating speed, is desalinating. This desalination is caused by the exposure of the saltwater to the polar ice. As this saltwater cools in this region, it releases some of its salt content. As discussed previously, this salt drops straight down to create a salt flat.

The salt, that has dropped out, is now lost to the conveyor belt. No longer is it there to facilitate the continued run of the belt that has enabled our temperate environment. Add to that the fact that the caps themselves are melting and further diluting the ocean's saltwater. The more it dilutes, the slower the movement of this life preserving Oceanic Conveyor Belt.

The slower the belt moves, the warmer the water at the equator becomes. Hence - Global Warming. With warmer water comes more water vapor. Increased water vapor compounds the problem by acting as a greenhouse gas. Now, the warming really begins to run amok.

More and more fresh water is on the move. Falling on lands that fail to add more salt to the process to maintain the balance. But, worst yet; an increased amount of snow fall will begin drop onto the much colder polar caps that was discussed in the chapter on the Oceanic Conveyor Belt. Therefore, in this scenario the polar caps will grow with the increase in the equatorial temperatures. This will begin to happen; if it hasn't already. Shedding ice can be a sign of growth. As the height of the ice increases, its weight will push down harder and squeeze other ice out the side. So calving does not mean the polar caps are shrinking, Yet!

As the salt content decreases, the equatorial temperatures increase, and the Oceanic Conveyor Belt slows; the polar caps begin to grow at a fantastic rate. Growing quickly both upward and outward. This growth will not be evenly distributed.

The northern cap having mostly saltwater underneath will shed weight faster than the southern cap. The southern cap having land under its ice will grow to a massive hunk of civilization destroying weight. Because the land mass under the southern cap is not perfectly centered on the south rotational pole, the destructive process will actually happen before the height of the ice surpasses the rise of the equatorial bulge. The equatorial bulge is the difference between the distance around the planet from north to south versus that of at the equator. The difference is approximately 13 miles. This additional material is the reason that the planet's crust has not shifted already. This is centrifugal force at work.

However, the weight spinning off balance will eventually pull the crust suddenly and violently as again centrifugal force takes over. In short, the polar caps will succumb to the forces of nature and move towards the equator. Life as we know it ends.

Do we have to accept this?

Is there something that can be done?

Do we have the technology to accomplish the civilization saving procedure?

The answer to the first question is a simple - no, we don't have to accept this. There is something we can do. The technology exists to prevent the destruction of our civilization. Many of science's greatest minds realize the problem of desalination; yet fail to identify the cause and thereby the possible preventive measures that we can take.

We need to bring together the powerful minds of our collective and have them calculate the amount of salt needed to bring the Oceanic Conveyor Belt back to its optimum operating levels. Then, we can begin to farm the current salt flats for the required amounts.

Having the farmed salts, we can introduce it at the equator by passing seawater around it. In essence, we create a small river or series of rivers to do what nature is no longer doing sufficiently.

As we gain knowledge of the correct amount of salt, we could form manmade rivers from the salt flats to the sea.

This would reduce the cost of our continued civilization's survival.

Eventually, after a great many years, those deposits (salt flats) will become depleted. The problem would return unless we develop a way to farm salt from the sea floor beneath the polar caps.

Simply churning the salt at the sea floor will probably not be sufficient to maintain the proper operation over the long term of many, many millennium. The reason is that the seawater that the churning salt comes in contact with will still be at a temperature that is holding all the salt it can through saturation. Thus, to add some sort of mixing device will only succeed in moving the solid salt haphazardly about. These much needed deposits will become spread about and unreachable.

Future salt farming will require us to bring the salt from the ocean's bottom, below the polar caps and reintroduce it in a controlled manner. As the process continues the salt will precipitate back into the farmed areas. A life preserving cycle will be started.

The threat of a crustal shift that is caused by natural global warming will be averted. Our intellectual powers will be able to look for any other triggers that may become the apparent demise of our future civilization. I say this because nature acts in cycles and we may interrupt one cyclic process; but, another will probably develop.

Chapter Ten

The Dead Sea

How It Died

The Dead Sea is not actually devoid of life. It has an ecology of its own. However, we call it the Dead Sea because it does not have life that we have become accustomed to finding in a body of water. That is due to the extremely high salt content.

The salinity of this sea is so high that fish, snails, clams, and the like are unable to live there. So, I gave this some thought. How did the Dead Sea get so salty? Is there an explanation that could be modeled in its creation? Are there other examples that could be used in comparison?

I saw the Dead Sea as an important piece of the puzzle that I was putting together. I did not understand, at first, why it was an item that needed to be solved. It was there and I just looked at it that way for a very long time. As a matter of fact, the implication of the Dead Sea did not dawn on me until I was already writing this book.

Not having an answer for this phenomena was causing me some concern. I believe that a theory that doesn't use all of the relevant and available clues (evidence) is probably not correct. So, I continued to mull over the Dead Sea. Then I became aware that the Black Sea was at one time, fresh water. That was my big "duh" moment. I began to reason that the Dead Sea was probably a fresh water body in the distant past.

This is another of my theories that I do not have the resources to prove. The ability to draw core samples seems easy; but, we all know that money drives everything and the lack of capital will leave a theory to the pages of a book or magazine, at best, or in conversation only, most likely.

I hope that someone will undertake the project to see if the Dead Sea was at one time a fresh water lake, if you will. I believe that those investigators will find that way back in history, the Dead Sea was alive with marine life that would thrive in fresh water. I, further, believe that this body of water evolved into a traditional saltwater sea and that the marine life had changed to survive in the new environment. Now that I explained my theory, I have to tell why it is important.

My first thought on the Dead Sea's creation was that the body of water formed over an ancient salt flat. This would cause the high salt levels and would support the theory that the planet's outer crust has shifted. The accumulation of water over a salt flat could create the

Dead Sea. The fact that the flat is where it is would add credence to the shifting crust. However, I did not think this was the way it happened.

Water accumulating over a salt flat would have instantly created the Dead Sea as the water came in contact with the salt. But, I felt other anomalies would have been created that would have probably stood out. First, I thought the salt flat would protect itself from the water to a certain point. The water would penetrate only so deep into the salt and then stop. Second, in order to turn into the Dead Sea, you would have to have a large amount of agitation to gain a mixing effect. I don't know if this is possible, so I rejected the idea.

I then envisioned a body of water forming as a fresh water mass then slowly, over a large amount of time, turning into a normal saltwater sea. Then as time continued to pass, an eventual "over-salting" to the point that the Dead Sea is now at has transpired. How this would happen is similar to the process described in previous chapters. Here are some highlights.

After the crust stabilized following a previous flip, a low elevation spot (this being the lowest) begins to hold water delivered by the rivers and streams. These rivers and streams carry very minute amounts of salt to this low spot. A lake/sea is now beginning to form. With the percentage of salt being so low, a fresh water accumulation is formed. This lake is subsequently subject to evaporation. The evaporation, of course, does not remove the salt from the location. Water is removed to become rain elsewhere as the salt remains. So, as rivers continue to deliver water with traces of salt and again evaporates, over and over again until the Dead Sea forms.

The reason that desalination does not occur is due to the fact that the Dead Sea does not have access to other bodies of water, such as an ocean. As a result, the water located in the Dead Sea does not come in contact with the polar regions of the earth. Without bringing the water's temperature to the point of near freezing, you do not have the process step nature uses to desalinate.

If the water remains warm, even hot, the water is able to hold the salt in suspension. Should this water become extremely cold, it would release the salt in the precipitation process step. As we know, the Dead Sea has no way in nature to become that cold.

It's at this point that I would like to remind you of the saltwater experiment that you may have performed with two bowls exposed to different environments. The one outside in the freezing temperatures or in your freezer had salt on its bottom in the form of a flat; but, showed no obvious evidence of salt creeping over its side. The bowl left exposed on the kitchen counter failed to form a salt flat as it evaporated; but, it did have salt creep up and over the sides of the bowl.

When you look at the shoreline and surrounding areas

around the Dead Sea, you are immediately struck by the fact that there is salt everywhere. Salt has attempted to hitchhike its way out of the lower elevations with the water as it evaporated. Just like the bowl that sat exposed on a kitchen counter. This emphasizes that a saltwater body of ancient times that dries up can not create a salt flat. The current scientific dogma on the subject is wrong. That is why I knew in the back of my head that the Dead Sea was important.

The next problem that would need to be looked at is why are the Dead Sea, Black Sea, Salt Lake, and others submitting to this process and other bodies of water around the planet are not?

The answer is simply that the rivers and streams that feed the other bodies are not coming in contact with salt deposits. By not coming in contact with salt, the flowing water can not deliver it. Salt is not everywhere and, more importantly for the replenishment of the oceans, is not in a never-ending supply. If rivers and streams do not continue to deliver new supplies of salt, yet are still providing water, the oceans will desalinate in a dramatic reversal of the process that converted fresh water bodies to saltwater. This is simply something that we cannot afford to have happen.

As I said in the beginning of this chapter - if a piece of evidence is being ignored or unanswered, then the theory is probably incorrect. This does not now apply to my hypothesis outlined in this book. I have put to rest the nagging thoughts concerning the Dead Sea.

Chapter Eleven

Catastrophe Averted

Future Threats

Let's say that we managed to overcome the cyclic powers of crustal shift, what now?

Earth Changes or the current buzzword "Global Warming", are indeed the cause of previous shifts and we have discussed how it happened through desalination of the sea. In a hypothetical future, we have averted the natural process. We need to now focus on removing the human threat to our planet.

I am obviously talking about our use of fossil fuels. Oil, natural gas, and coal introduce hydrocarbons into our atmosphere when we burn them. These hydrocarbons then become greenhouse gases that would continue to threaten our civilization's existence.

If we continue to burn these substances to power our world, we will eventually fuel the cyclic process that will result in a crustal shift. Salt farming will only buy us some time. We have to find a way to run our civilization.

Nuclear power plants have their unique hazards. They would only succeed in giving us a new major problem to deal with.

Is hydrogen the answer?

No - hydrogen, when used as a power source through combustion, produces water vapor. Water vapor is also a greenhouse gas. It is also fresh water, which is a threat to desalination.

In addition, imagine all combustion-engine driven

devices emitting water vapor. Every car, bus, train, plane, and countless other devices we require introducing water vapor into the atmosphere. My first thought on the matter was to envision a city where it rains all of the time as the vapor condenses by coming in contact with relatively cool buildings that reach to the sky. But that problem paled in comparison to the global effects that would be set into motion.

Global Warming would still be a threat and we would be speeding it up. Our attempt at stopping Global Warming would be for not. We need to find a power source that does not require the burning of anything. We need to not introduce anything into our atmosphere. Anything we add will probably lead to unexpected results. Nature will eventually use what was given it to return to the cycle. After all, there are three other natural causes that give the impression of Global Warming.

They are: An increase in the number of solar flares on

the sun; the change in our angle of our planetary tilt to a more upright position; and a phenomenon I call "star wobble" which is a counter-pulling effect of large planets rotating around a star. The observation of this star wobble is used to locate other stars with planets. A star moves toward its large planets from our point of view allowing us to know that a large object is there despite the fact that we cannot actually see. I would now point out that our star has some large bodies around it and they are beginning to align on one side of solar system. This would affect the distance between earth and the sun; thus, warmer temperatures.

We really have no power over these types of events. One natural process seems to be the only one we can change. So, unless we shut down our civilization, we need to find a new source of power.

I am a realist. I am aware that any new power source will still need to be controlled. The power companies will still need to charge the end user. I am not a nut, who believes in something for nothing. I understand the economics of this civilization we live in. Money will still need to trade hands - that's a given.

Cold fusion appears to be years away. That is, if it is even possible. Perpetual machines are unlikely, due to the effects of friction. So, we have a dilemma.

I propose a possible answer. This answer has been looked at before; but, has been grounded to this planet. It is solar power. The collection of solar power has been proven both here on the ground; which is limited because of the day/night process and weather, and on spacecraft/satellites. It works!!

We need to think outside of our atmosphere in order to have a continuous flow of power. Above the effects of the planet's rotation and weather patterns, we find a never ending source of power in the sun.

The trick is to collect it, get it to the surface of the

planet, and subsequently distribute it to the paying end user. Challenges are there when you look at this proposal. The size of the collectors and their mechanical make up is but the first. I don't have the intellectual power to tell you the answer - but, I know our collective can overcome this hurdle.

The next challenge is how to get the power down to earth. To me the answer seems simple on the surface. We need to transmit it back to us. After all - radio and television are the broadcasting of a modulated electrical signal. Up size the transmission to ground based receivers, owned by the power companies. These companies could finance the project, maintain it, and continue to distribute the power. It would be a great investment based on the Return on Investment (ROI). Once it is up and running the overhead will reduced to zero on the actual power generation. The balance of our economic society remains in place.

The final challenge would be to get the power to the end user - you. To that, I state that our current hard wires would still be usable and a system of broadcast could be set up for the mobile power consumers.

I believe that power broadcasting is something we have the capability to do since the time of Nicola Tesla (early 1900's), the grand master of electrical power use to date. Nearly everything that runs on electricity today owes its birth to the intellectual prowess of Tesla.

Anyway, reports and writings of Tesla's work suggest that he was able to broadcast large amounts of power. On a much smaller scale, radio and TV confirm we can broadcast electrical power. It was probably shelved years ago because a means to bill the end user did not exist. With no way to charge us for the amount used - the power companies had no reason to invest in it. Today though, we can be monitored for our usage.

Every time you call someone with your cell phone, you

connect with a system that knows you are using it and for how long. It even keeps the signal open to you and your conversation's partner. Point to point and, most importantly, controlled.

If we bring this technology to bear on our problem, we would end up with an electrically powered automobile that is running on broadcast power, through a connection similar to that of a cell phone. If you have the key on - you are being billed. As you increase your power usage by stepping on the gas - if you will - your bill from the power company will go up. The age of cell phones has facilitated this system of power distribution.

We, as a collective, have the technology, the oil, power, and automotive industry have the financial resources, and, more importantly, our civilization has the need. If we fail - we (civilization) cease to exist as we know it. Can we afford to not do something?

Chapter Twelve

A Hard Life After

What We Can Expect

I often imagine the "What if's" that would be associated with a failure to prevent the next flip of our planet's crust. Imagine if you will, what life would be like immediately before, during, and right after the crustal shift. If you have not envisioned hell on earth, then you haven't taken enough factors into consideration. You must think of the extreme weather patterns and changes, earthquakes, volcanoes, tsunamis, and the sudden change in the earth's rotation just to name a few of the natural influences. Then, if you can handle the horrors of all that, add the items that mankind will contribute. Any survivors will have their work cut out for them.

As we get closer to the actual event, we will be affected by the major weather changes. Rains in areas that will not be able to deal with all of the water will become flooded. Droughts will begin in areas that are normally fertile. Areas that are already dry will become unbearably hot. Wind directions will change and the wind speeds will increase. Areas of major temperature differences will suffer from tornadoes, typhoons, and hurricanes at colossal levels. Temperatures will increase in some locations, while they will decrease in others. The ability to bring crops and livestock to market will become more and more difficult, if not impossible.

Then as we get closer to the main event, the stresses and strains on the tectonic plates will increase. This will happen when the Oceanic Conveyor nears the point of shutting down. The equatorial temperatures will rise, possibly to the point of boiling. The evaporation rate will be at such a rate that steam will envelop the entire globe.

Where the water vapor comes in contact with cold air, around the Polar Regions, snow will fall constantly. This unending snowfall will accumulate in an area that is no longer receiving heat from the ocean in the form of thermo transfer. The Polar Regions, whose outer boundaries have increased, will grow at a staggering rate. As the snow accumulates, the weight will increase to form ice that will give future generations the impression of an ice age.

While the weight of this ice increases, it will put additional stress and strain on the tectonic plates that normally would not be there. The plates surrounding the Polar Regions already have a stress level higher than that of the plates near the equator. This is because the drag on the outer crust that results from the core spinning below is less than that created at the equator. There is less contact area under the spinning poles than there is at the equator. Thus, there is more friction/drag to cause the crust to follow the core's lead at the equator.

In essence, the tectonic plates along the equator are pulling the polar plates around in the earth's rotation. With less friction/drag comparatively, the polar caps are influenced at a greater rate to the physics of a body at rest wanting to remain at rest. If the tectonic plates were not in contact with each other, the equatorial plates would probably rotate faster than the polar plates in a ratio of 3 to 2. For 3 equatorial rotations the polar region would rotate 2 times.

So, the crust is not trying to move together - it is being forced to. As mentioned above, if you took the "pull along" force from the equatorial plates away from the polar plates - making them free floating - you would end up with differing rotational speeds of the outer crust.

Think of it as you see our jet stream winds. The winds travel from west to east as our atmosphere tries to keep up with the cores rotational speed. This desire to rotate at different speeds produces great stress and strain on the tectonic plate's edges.

With this stress and strain, add a major increase in weight that comes from the accumulation of ice. This downward force will need to be compensated for elsewhere. A relief of pressure will take place somewhere. As you can imagine, this will come in the form of earthquakes. These quakes will be exhibited in areas that are normally stable. These stable areas will begin to move because the normally fluid plates will not be able to relieve enough pressure. We will be confronted with earthquakes in places that we didn't even know there were faults.

Now, with the crust on the move at rates not normally seen, we will begin to see more volcanoes. Areas in the plate that had been holding magma in check will succumb. This will be caused by the increased amounts of water that will end up underneath the tectonic plates. The openings will shift and move so much that excessive volumes of water will become trapped under the plates and in contact with the magma. As the water boils and vaporizes, the pressure will build to levels beyond the abilities of the vents and geysers pressure relief. The pressure from this vapor will eventually build to the point that earth is unable to hold.

This will blow holes in the crust where the path of least resistance is located in the crust and allow magma to reach the surface. We now have lava to deal with. A situation that will be apparent in a many places. Starting first in traditional volcanoes, then in areas with relatively small amounts of resistance; such as Yellowstone, and then, if the pressure is great enough, in places where it appears that volcanic activity has been non-existent for eons; for instance, Taylor's Falls, Minnesota.

Remember that water at super heated temperatures will create the thrusting power that will cause an eruption. This is why normally volcanoes are displayed in the form of mountain tops. The steam rises up from the magma into the throat of the mountain and becomes trapped. As more pressure builds it eventually achieves the force needed to blow the rocks out of the way. But, there will be unimaginable amounts of water under the plates and the need for relief will overwhelm the traditional volcanoes. Thus, new volcanoes will sprout up everywhere.

Moving on to another issue - with the crust virtually floating free and an overwhelming weight sitting on the polar caps - we are faced with a major physics problem. The equatorial bulge, if it still exists, will be unable to counter centrifugal force. We have now reached the event horizon. The flip begins.

The influences of centrifugal force are now in play. The massive amounts of snow and ice that have accumulated at the poles will succumb to the natural influence. The poles will have, with the help of the equatorial bulge, fought to remain balanced at the earth's pivot points for as long as they could. However, because they are not perfectly centered on those points, they surrender. The poles, as we know them, will shift their positions to the outer edges of our circular (rotational) motion. In other words, the caps will move toward to the equator.

I have speculated as to the direction of travel that the ice caps will take. I have imagined the maximum speed of the shift. I have even theorized on where relative safe zones will be located. So, you may ask where are the places that will survive? I am not going to provide my thoughts on this matter for numerous reasons.

First of which is that we need to do something now to prevent the event and I believe that resources would be wasted trying to save some people and sacrifice others. This would be understandable if there were not a promising procedure available to prevent the destruction of our society. Second - if I were to be incorrect in my calculations and mankind "bet the house" on my chosen position on the planet, we could end up extinct. That is more than I want to claim responsibility for. I am sure others will begin to find their own safe zones. So I will leave it at that. The last reason I am going to save my ideas on the topic is - if I am right and millions of people gather at these points, they may over populate the available resources there. Life after the crustal shift will be hard enough, as I will describe further, without having too many people in one place that will fight each other because of their primal instinct to survive. As cold as it might seem - we are better off as a species to stay right where we are and take our chances. That is what my family and I are going to do.

With that having been said - let's get back to the imaginary (for now) life during the flip.

You need to know that there can be no more energy exerted than there is stored within. In other words, I am saying that the maximum speed of the flip at any one spot will not exceed the rotational speed of the planet. We will be limited to approximately 1100 miles per hour when the shift performs its act. This maximum shifting speed will not be survivable for those located along the direct shifting path for obvious reasons; but, it will be important in determining the events that will transpire for those who are in other areas.

When the shift starts, the crust will, in relation to north and south movement, as we feel it, go from zero to 1100 miles per hour and then back to zero. The ice is stationary now and will be again when the ice masses reach the new equator. The acceleration and subsequent deceleration will be life destroying violent along the direct line of travel; but, less so as the positions move away from this line of travel.

Thanks to our current rotation, the movement will not be on a direct 90 degree heading. This is because as an area begins to move north or south, it will still be moving east. This will create a swooping curve movement. This swooping movement will be less violent the farther you are from the direct line of travel. People located 90 degrees away from the direct line of travel may not even notice the crustal shift as far as the actual movement is concerned. Other hints to the shift will be noticeable though.

Have you ever seen a spinning top eventually slow down and fall over? Remember the wild movements of that toy? While this is not an exact representation, it can give you an idea of the events that will happen.

Imagine being on one spot of the spinning top as it falls over. First, you are moving one direction in the rotation, and then you suddenly move in a different direction. Then as the top settles into its new position, it moves (rotates) on a completely new axis. In essence, it has a new equator. Needless to say, the change is violent. Yet, if you slow the imaginary top down to a speed that can be looked at frame by frame, you will see areas that are less affected by the toppling of the top.

With your imagination, using a frame by frame look at the top, envision a spot that is along the direct line of directional change. You will notice a sharp, well defined change in direction. Yet, if you look at a position that is 90 degrees off the direct line and you will see a sweeping, relatively slow change in direction. This position will virtually not be effected by the return to zero speed because it will glide into the new equatorial rotation with much less violence than other areas. This less violent environment exampled in this analogy is the areas on this planet that will hold the hopes for mankind's continued existence. If we are to survive as a species, it will be in these areas. But, it won't be easy. Other destructive events will be working to try and snuff us out.

Water on the move will be a major problem. Tsunamis, tidal waves, and other walls of water will be everywhere. These massive mountains of water will be crushing everything in their paths. Where a body of water exists – there will be destruction as far as it can reach. This reach will be extended to points beyond our imagination.

Remember carrying a full glass across a room. Traveling along, balancing your hand so that none of the liquid is carried over the rim of the glass. The kinetic energy of the liquid is trying to stay where you were, while you are working to move it from one spot to another. Can you also remember what happened when you were carrying your drink across the room and someone bumped into you?

The sudden change in direction results in the liquid

rising up and over the rim of the glass. We call it splashing. This splashing effect is in the opposite direction of the glass' movement. If the glass moves rapidly to the left, the fluid splashes over the right side. Again, the kinetic energy is trying to keep the liquid in the spot it just vacated.

This type of liquid movement will take place during the crustal shift. Water that is sitting relatively stagnant, will try to remain so as the earth rotation changes direction suddenly. With no energy being expended on bodies of water in a north/south movement, there will be no instant acceptance of this sudden movement. As the earth's crust shifts north, the water will give the impression of a southbound movement. Tsunamis/tidal waves will travel, from our current point of view, south and destroy everything in its path for as far as it can reach. The Tsunami of 2000 CE that took place along the coastal regions of the northern Indian Ocean will pale in comparison.

With the first volley of tsunamis/tidal waves having conducted its destruction in a direction opposite the line of crustal shift, you might think that the worries are over. Wrong! Just like the numerous huge waves caused by earthquakes prior to the shift, we can expect them to continue long after. After all, the forces that were in play causing immense tension on the fault lines will be a long drawn out event. These earthquakes and tsunamis/tidal waves will run rampant until the earth settles into its new rotation. A rotation that won't change for the planet's core; but, will appear completely different on the surface. The sun will now appear to rise in the south and set in the north or vice versa. It will take awhile for the crust to find a smooth transition to this.

The first of the new problems will be the introduction of the polar ice to the new equator. All of this ice will be melting at an astronomical rate. A rate that will exceed nature's ability to return to ice-form at the new poles. This will cause the oceanic water levels to rise dramatically. If Greenland's ice were to melt alone, the water level will rise about 23 feet. It is believed that the polar ice melt would be 11 times as much. So, sea levels will probably rise 250 feet or more as a result of the shift.

Now try to imagine the unorthodox weather that will be taking place before, during, and after this event. When the Oceanic Conveyor Belt begins to shut down, the equatorial temperatures will rise. The hotter this region gets, the more water vapor there will be in the atmosphere compounding global warming. This increased vapor has weight to it. As a result the air will get heavier and will lag behind the rotational speed of the earth's crust. Weather patterns that move generally eastward will drop back and begin to move to the west. Just like a hurricane that is loaded with water, moves west until it releases the vapor in the form of rain. After it lightens its burden, the hurricane makes a turn and heads back to the east following the prevailing winds.

This continued change in the weather patterns will disrupt our abilities to produce crops. A failure to grow food will make for a great many starvation deaths and will lead to subsequent hostilities between those that "have" and those that "have not". A "Mad Max" scenario will be prevalent. Further, this inability to grow crops will be complicated by the fact that because of the unpredictability of the crustal shift, we won't even know what season it is. Despite the fact it may have been winter prior to the shift, it may become summer for that hemisphere immediately after. This will be due to the positional change with regard to our planetary tilt. Therefore, a Stonehenge will need to be constructed, after the ground stabilizes, to determine the new seasonal pattern and your exact position on earth with regard to latitude and longitude.

As if these natural problems are not enough, mankind will also work against you as a survivor. I mentioned the "Mad Max" scenario above and that will be brutal; however, other survivors are just part of the new problems you will face. Disease from the rotting flesh of those that did not survive will be everywhere and unavoidable. You will have a full time job trying to bury or burn the immense numbers of dead. An estimated 300,000 died in the 2000CE Asian tsunami alone. This number will be considered minuscule in comparison.

And our civilization from before the crustal shift will still be there trying to finish you off. A literal "blast from the past". This will come in the form of nuclear power plants that are continuing to react without any of the control measures in place. That's bad - right? Yes and so is a nuclear bomb that is now left to its own devices. Like a landmine waiting for its intended purpose. Speaking of landmines - what about them in the next epoch?

Hell on earth may be an understatement. Living to the next epoch may not qualify as being a survivor. The odds will be greatly stacked against you and, on a grander scale, against mankind as a species. If we continue to ignore the problem some of us will find out just what nature has in store for us. See you there - MAYBE!

Chapter Thirteen

Years Later

What We Look Like To Them

Assuming that we fail to act and the destructive power of the crustal shift is brought to bear, what would our past civilization look like years from the event?

Imagine that the crust has shifted. We have managed to survive as a species. There are bands of people scattered about on the land masses. Some are on what is left of the continents as we knew them and some are on any newly inhabitable bodies.

The earth has stabilized after a number of settling

shifts. These shifts were caused by the buildup of ice on the new polar caps as the old caps melted down. The earthquakes have subsided and the volcanoes have relieved enough steam pressure to return to their dormant state. The sea level has achieved its maximum height and no longer threatens to drown people and animals.

A new Oceanic Conveyor Belt is establishing itself and is now beginning to create stable weather patterns. Our existing weather-persons are becoming more accurate in their predictions of systems. The seasons have been normalized despite our inability to identify the calendar date. Planting can commence once we know when spring starts.

Our knowledge of crop growing has been passed from one generation to the next. After all, this is very important information. To this point, we, the survivors, have been hunting and gathering for our subsistence. We have been wandering about for years, like nomads, in search of a location to settle on.

First priority now is to determine the date and our exact position with regard to latitude and longitude. We build a Stonehenge. A structure that takes a minimum of a year to bring to completion. With this device, we can determine the start of spring for planting and based on the interpretation of the rising of the sun, we can also find our location on the planet.

This location is not very important with regard to comparing it to our memory of the globe, things will have changed dramatically. It is extremely important in determining if we are located in a region that may become overcome by extreme weather. We don't want to remain in a new desert that will receive little rain and we shouldn't stay in a new rainforest and try to plant corn. It won't work - too much rain.

So, yet again, we may be moving. We, "the lucky

survivors", live a mobile life for many generations. Constantly trying to change our location, in order to find the best possible land to live on. Many of us are unable to move to an area that meets our perfect desires. We end up finding natural barriers that prohibit our movement.

A new coastline to a great sea is but one such barrier that we may come upon. This, however, is a blessing. With the sea, we are able to provide for our needs. Fishing will be one of our best opportunities for sustaining life. While much of the aquatic life has suffered from the devastation, the largest survival rate for animals will be in the sea. Food will be available there.

Another benefit to life near the sea is the ability to travel. We will be able to build rafts and rudimentary boats to travel at a quicker pace to happier hunting grounds. The start of new lines of communication will happen along the coasts of the new continents.

Continued repopulation will require the finding of outside bloodlines to procreate with. And most importantly for those that end up on islands with limited food supplies, a need to set sea for larger landmasses will be paramount.

During all of this movement in search of a relative paradise, we have failed to commit past knowledge to a permanent means of retrieval. The internet is gone and books are too heavy to transport. Thus, many things that we are able to call upon are gone. Just think of the things you know now, that will become virtually useless in the next epoch because it won't help you to survive.

Teaching your off-spring to hunt, fish, and cultivate will be the number one things that can be taught them. Nuclear physics, geology, and archeology would not last long in our knowledge banks. Continued survival will be all important. Reading and writing abilities will suffer greatly. Things that are passed along verbally will become distorted. Many of us in school have performed the little experiment where the teacher relates a statement to one student who passes it to the next and so on until the last student tells what he or she heard. Remember, that the last student told a story that was vastly different from the original sentence. Think of what that will do to our knowledge base. A great number of subjects will be forgotten or warped. A "dumbing-down" of our civilization has and will continue to take place.

Eventually, even the story of the crustal shift will be turned into a myth or legend, if not completely forgotten. Some skills will be passed in a verbal state; but, the ability to perform the tasks will not be available.

The building of skyscrapers may be remembered; but, we will be unable to do it. The machinery that we have come to rely on will have long ran out of gas and/or fell in disrepair. Soon, they will rust away. Our need for these types of equipment will have been supplanted with other more immediate requirements. Much of our current knowledge will become just worthless trivia.

In some areas, vital knowledge will thrive while it fades away in others. If a metal worker survives at one location and he will pass the trade on. A group of people will be able to make tools provided they can meet the technological requirements to build forges. In other places, this expertise dies and you will have to resort to making tools out of bones and stones.

Just think of the things we take for granted. Tools are just one. Cookware will eventually have to be fashioned out of the available resources. Weapons will de-evolve from firearms to spears, bows and arrows, and knives. Even making a wheel will become a major feat. Little things will be nearly impossible to fashion; fish hooks for instance. Ropes will not be at the local hardware store. You will have to be able to fashion one yourself. Have you ever made a fishing-net with the things found in your yard?

By now, you should be able to envision just how backwards the early generations following the crustal shift will appear to our future, intellectually powerful descendants. Following many, many years, the people of the next epoch will evolve their knowledge back to the level comparable to ours. They will have machinery of all types. They will have the power of flight. Everything we do now, will again be performed. It will just take the creation of a written language that can record the lessons learned on the intellectual journey back to advanced technology. With a written language, people will not be required to memorize things that are important to progress. They will be able to refer to this knowledge as needed. Just like today, if you can read you will have the ability to do things that are not normally in your abilities.

With the importance of knowledge in written form having been addressed, you might say that we need to try and build a library now in some sort of hardened structure. This library's purpose would be to pass knowledge to the next epoch. Don't waste your time!

Electronic storage devices will be useless without electricity or the software to interpret it. Film will deteriorate. Books will rot, if they don't burn. Worst yet, our languages will evolve. The English word may still be spoken; but, who will be around to teach the masses to read it. A language of pictographs will be misinterpreted and the translation will be only as accurate as the reader. A reader who may believe that the ancients (us) were not capable of things we now take for granted. After all, how sure are we of the translation of hieroglyphics. Think about it!

Now that we are talking about the people of a few thousand years following the crustal shift, let's look at what they would see when they looked at our epoch. Their archeologists will be digging up some remnants of our civilization and their geologists will be working to determine the meaning of their findings. The future paleontologists will be reconstructing the fossils they find. Things will be just as they are now. These scientists will be ignoring myths and legends, assuming that their epoch has achieved the highest intellectual level to date. This will be an assumption that will be justified without a coherent written record of events to tell them different. I wouldn't blame them, just as I don't blame our current scientists.

They will be formulating theories based on the available evidence that they have. Evidence that will be slow in coming to them. Their early explorers and treasure hunters will find items presenting small pieces of the puzzle. Starting with fossils that will still date back to long before us and every now and then, they will find something left behind by us. Insignificant items that hint at who we were. Nothing earthshaking will remain to be found. The upheaval will have wiped out anything that really identifies us.

They will be finding the very things that we find now. The apparent ice age of our time will be the basis for their ice age theory. Wherever our polar caps end up will leave the tale tell signs that they are correct. The evidence we use as a collective to support our theory of an ice age will still be apparent. This will encourage their belief that ice ages are a recurring event. They will interpret the evidence as the cause of the strata layers, disregarding the same evidence we do today.

Eventually, they will find man or animals caught in the permafrost. They too, will probably think of this as a group that lives in the cold. Ignoring, the fact, that these life forms were more appropriately warm weather animals. No notice will be given to their instinct for survival that should have caused them to move away from danger. Our scientists do just that, so why wouldn't future scientists?

Mass extinctions will be evident. Large numbers of

animal life will not be present in the next epoch. This will be most common among animals that are only found in small regions of our present earthly layout. An example could be polar bears that, for obvious reasons will be threatened with complete annihilation.

As their scientists' abilities expand, they will become able to draw core samples from beneath their own polar caps. They will then find that at one time, lush greenery was growing at that location. Strangely enough, they will find the same thing we did. The flora will still be green. Not brown and shriveled up, like it should be if the process of freezing was a gradual occurrence. This finding will be the future scientists' first indication that they are wrong in their belief. Suddenly, they will review other anomalies to find a new theory that won't go too far out on the lunatic fringe.

They will reexamine at the wandering magnetic poles, strata layer formation, fossil creation, desert core samples, and the locations of bones and fossils from antiquity that seem misplaced.

Hopefully, it won't take an average person like my-self to try and draw attention to the implications of this evidence. I hope that their scientific base will not be driven or controlled by a system of funds doled out by grant. I believe that scientists can handle debate. They can argue their theories and take a risk with expressing their thoughts. I don't, however, believe these same scientists can do that and watch their grant money dry up. In our next epoch, if it is left to happen, we will need to create a means of funding the exploration of thoughts that won't be controlled by organizations of the government or entities that receive capital from the government.

Assuming that this dilemma has been dealt with in our future, the scientists will examine the wandering of the magnetic poles. They will determine that small deviations can be expected. The earth is a dynamo. A dynamo that is creating a magnetic field encompassing the entire planet. This field is a result of electrical power generation with positive and negative terminals, if you will. What is different from other dynamos that we have become accustom to is that there are tons and tons of material covering it. Some of this material has differing resistive properties that work to block the path of electricity to the atmosphere. This power does eventually find a track to follow and it exits the surface of the planet. Every now and then, the exit point moves a little due to tectonic movement that changes the path of least resistance. But, this won't explain large changes in the magnetic poles. Changes that are better determined by a crustal shift. If the core remains in its position and the crust slides to a new orientation, a dramatic movement in the magnetic pole will result.

This shifting will then explain with ease, the formation

of strata layers. A sudden environmental change will have taken place in just about every location on the planet. Areas of lush, green flora could become desert. Desert could become ocean bottoms. Even new mountain regions could be created. The limitations for strata layer formation when looking at ice ages would still be present; yet, someone would still have to come up with a viable theory for an ice age to occur. A problem that has not been answered now and will not be solved by traditional meteor impact theories in the future. Don't forget, in order to get snow to form the ice, you need heat. A meteor impact or volcanic activity throwing debris into the atmosphere will only block the sun's rays for a relatively short time. If this happened to the point of causing a dramatic drop in temperature, everything would have frozen in place. Therefore, water frozen at the equator will not be able to make the trek to the poles to cause glaciations. This would have prevented the creation of over-sized polar caps. But, if you look at the spots that were former caps, you will see the results of mobile ice that cause spectacular rock formations and the effects of caused by the immense weight of the ice.

Our future scientists will already be confronted with trying to explain the odd locations of bones and fossils. Whales found in mountain ranges will definitely be a hard to answer anomaly that will probably be attributed to an gigantic tidal wave carrying the animal to its resting place. If this is how it happened, what would have caused a wave of that size? Remember cause and effect. The cause could have been the wave and the whale in the mountains could be the effect. We (they) are faced with a new question - What was the cause that effected the tidal wave? What type of natural event could result in a wave that could carry an animal weighing tons up into the mountains, thousands of feet above sea level?

If that isn't enough of an oddity - How was the effects

of oxidation neutralized and what kept the scavengers at bay? Why are the bones and fossils even there? You would need a colossal event to leave this type of evidence. The fact that fossils are formed and bones remain for anyone to dig up creates an enigma that is not answered. An average Joe like myself, can not understand any other way this could have happened. I will concede that a meteor impact could create the situation that killed dinosaurs and sufficiently protected their bones; but, that would not explain all of the other examples of bones and fossils. One impact equals one strata layer - if the environment appears to return to its original climate, say from forest to desert and then back to forest. Sprinkle in some iridium, if appropriate. Then I would lean towards an agreement with a meteor theory. Yet, this does not explain all of the others. Even more importantly, don't forget the animals caught in permafrost in other epochs that somehow ignored their survival instinct.

These are some very hard questions for our future scientists; but they will become easier to explain as their technology advances. They too, will be taking core samples of their deserts and polar caps. Do you think they will find anything different than we do? Of course not!

Under the desert they will find evidence of a more hospitable environments. Trees, jungles, savannas, who knows! What I know they will find is something vastly different from a desert - unless a desert was lying at the rotational point of the crustal shift. If this is found, core samples taken even deeper will still show what we found and other deserts in their epoch will display something similar to what we found at a shallower depth. If this situation arises, they will be given a tool to determine a timeframe between shifts. A bonus!

If you take anything away from this chapter - I hope it is the similarity with which we are finding the same evidence. The repeating of this cycle will forever create the same evidence. No changes in the examples will be found from strata layer to strata layer. Epoch to epoch will constantly provide us, if we survive, with the same pieces of the puzzle.

Following each event, mankind will go through a dumbing-down and subsequent rebuilding of intellectual prowess. How far our intellect will grow will be reliant on the amount of time between crustal shifts. Short terms in the interim will result in a group of people who seem relatively backwards. A timeframe comparable to ours will get results on par with our epoch, and of course, longer stretches between may result in a civilization far advanced to ours.

Is there hope for a future advanced society? Yes, there is, once they find a way to prevent the shift. After all, I have proposed one. However, there is another way. A period of time between shifts that is longer than ours will enable this type of intellectual growth. The possibility of an extended period between shifts lies in a world of chance. If the future generation is lucky enough, the planet will settle on its new polar caps with large land masses underneath.

Should this happen - the period between shifts will be drawn out due to the protection of the saltwater offered by the land. The land will prevent the saltwater from coming in contact with the polar ice. This will result in a slowing of the natural desalination of the oceans which is the overall cause of the event. This may not be ideal, when you consider the increased percentage of salt that will be present in the water; but, it will extend the culmination of the process.

Conversely, if the next epoch ends up with the polar caps located over bodies of saltwater, they will be faced with an accelerated process and therefore, a shorter time between shifts. An immediate desalination will be taking place right after the shift, at a rate faster than that of our process. Two polar caps that have contact with the sea water from day one will give the desalination process a jump start compared to that of our epoch.

We had the landmass at the south pole protecting the saltwater for a long time before the ice reached the sea. It becomes apparent that the intellectual growth of mankind is actually left to a roll of the dice. “Pure Chance”. Let’s hope that if the need arises, that the next epoch doesn’t come up “craps”.

Chapter Fourteen

The Reader's Tasks

What You Need To Do

At first glance - you might think this chapter will be a survival guide to be committed to memory. An attempt to impart knowledge to help you continue your existence following the shift. Not so! This chapter is written as a plea for your help.

Our continued existence may be reliant on your ability to scratch out a life on the other side of this cataclysmic event; but, that is not the primary purpose of this section. There are things that can be done now that have priority.

Then, if we fail, you will have another mission. First things first.

I am asking that you take a very serious look at this theory. Please review “The HAB Theory“ by Allan W. Eckert, “Cataclysms of the Earth“ by Hugh A. Brown, and “Path of the Pole“ by Charles H. Hapgood. These books are on similar subjects that use many of the same items of evidence. They come to much the same conclusion that I do. Brown, Eckert, and Hapgood have given this cyclic event some serious thought. Einstein has given a positive endorsement to the plausibility of the theory. I have spent over 20 years looking at it.

I believe that this has happened. I believe that, unless we act as a global alliance, it will happen again. I have written this book in order to get your attention. I feel that this is the only thing that I can do by myself.

It’s now, your turn to act. The future of mankind and our civilization is in your hands. What do I want you to do?

First - I need you to doubt what I have written. Question everything that I expressed. Don't assume that I have it right. More importantly, don't automatically dismiss my thoughts. Investigate for yourself. Do some research. Try to prove me and the others listed above wrong. Bring to bear every type of anomaly you can think of. Get involved by using this topic as a brain puzzle. See if you can answer the things addressed in a different way. Reading the works of Brown, Eckert, and Hapgood will help. If you believe that there is a different answer - express it. Start a global debate on the issue. The internet will be beneficial in doing both the research and the debating. I don't care if I am wrong - if it means that mankind can continue as is.

Second - If, after you have done your own research, you become even slightly convinced that this theory is correct - you are asked to spread the word. Tell your friends, families, and acquaintances. Tell your local government officials. Tell your national representatives. And, if possible, tell any scientist you know. Tell Everyone. Yes, it may result in book sales for myself; but, that isn't the primary purpose for my plea. My request comes from the fact that information that, is not shared, will not enlighten the masses. Information that isn't passed on cannot be entered into the databanks of those that can facilitate its proper use. Information that is ignored can do nothing. We (you) cannot afford to let this topic be relegated to the shelves for books written by the lunatic fringe, even if you think I am crazy. The squeaky wheel gets the grease. You need to be the squeaky wheel. You have it in your power to make sure that the intellectual prowess of our collective can be brought to bear on this subject. Our civilization's continued existence is in doubt. The daily news confirms that every night.

Lastly, if we fail to prevent the destruction of our world through a crustal shift - I ask that you endeavor to help future scientists. This can be done by you placing a high priority on continuing to teach your off-spring to read and write while you struggle to survive. This will enable the creation of a written record of events that is accurately interpreted. This will be the biggest contribution you can make to the next generation of our species. A record and the ability to read it, will give the future humans a head start understanding what happened and, if possible, the tools to formulate their own attempts at preventing the event. Take away - now, the dumbing-down effect that will be inevitable on the other side by concentrating on reading and writing. Make sure that those that become your descendents know that the most important thing to learn is how to read and write. With this ability, you can learn about everything else, including what happened and why. If you need to, take with you and protect a copy of the dictionary. Treat it with respect that it will deserve. You might even take copies of books on the subject that relate what happened so that you can be sure the future scientists will empowered with a factual record. This will arm them with knowledge that will help to combat their own coming cataclysmic event.

You as a reader are not powerless. Your tasks have been outlined. They are not hard to accomplish. So, are you ready to save civilization?

ARE WE WORTH OUR SALT?

I truly hope so!!!

Afterword

As the process of writing this book comes to a close, a number of items that could have been included have come to light. They support the thesis outlined herein. However, I need to stop and publish at sometime. Continuing to look for evidence will only delay getting the message out. With that having been said, I feel it is important to briefly address some of the latest news.

First, recorded global temperature data is now suggesting that our overall climate is possibly stabilizing, if not - in fact - becoming colder. Many of those who have gone on record as supporting Global Warming are now switching gears and proclaiming that the beginnings of a new ice age are upon us. I feel that they are just now coming to understand the environment changes that I wrote about with regard to a desalination of the ocean and its effects on the Oceanic Conveyor.

These changes will eventually become even more evident. This is already happening. It now rains on the continent of Antarctica. The rain is caused by the rise in temperatures in the region that would normally remain cold enough to only allow for snow as the form of precipitation. This rain is the last thing that the Oceanic Conveyor needs. The rain falls as fresh water, softens the upper layers of snow, and; worst yet, gets down into the crevasses inside the ice and causes breakages. All of this rain water, softened snow, and free floating ice greatly affects the salt to water ratio at a much faster rate.

The second item of interest is "Dakota" the prehistoric dinosaur found in the United States. What makes Dakota special is the fact that he has skin included in his fossilization. This identifies that Dakota was quickly frozen in the transition from one epoch to another causing his death and protecting him from oxidation and predators. Then, as his remains again became exposed in a third epoch as the crust shifts for the second time in which his remains were present - just like will happen with the wooly mammoths frozen in permafrost, Dakota required another means to preserve his skin. It is my thought that this animal found itself encased in salt that formed under an ancient ice cap. A salt flat in the past. The salt would have essentially mummified Dakota and set into motion a system of preservation for his skin. I am quite sure that large quantities of salt were surrounding this fossil. If these salt deposits were not entirely depleted from past water run-off to the sea, we should be able to confirm this hypothesis.

The last item I wish to quickly cover, as I prepare for publication, is the fact that the Norwegian Government has taken steps in an attempt make food-plant seeds available to survivors of an upcoming catastrophic event.

They are storing vast amounts of seeds in the bunkers built to defend the coastline during World War II. These bunkers are huge. I have seen them and they will hold almost immeasurable amounts. But, that is not the extent of Norway's effort. They are further sending to sea large cargo ships loaded with seeds. I postulate that this is a backup plan in case the ground based seeds are destroyed. So, what event could cause the Norwegians to fear that any one system of seed preservation may not be enough? If I have done my job well, you already know.

About the Author

Arthur Ryan is a pseudonym for a man who wishes to maintain his privacy. An average person living in the Twin Cities area of Minnesota. Art's hobbies involve exploring theories for validity with the eyes of a detective or as a mechanic trying to diagnose a problem. He finds books on a variety of subjects enjoyable. His household is filled with family: Father - Ben, Wife - Wendy, Sons - Jesse and Joey, Nephew - Jake with his bride-to-be Shelly.

www.ingramcontent.com/pod-product-compliance
Lightning Source LLC
LaVergne TN
LVHW090947080826
845145LV00003B/914